Handbook of
Data Recording, Maintenance *and* Management *for the* Biomedical Sciences

Handbook of
Data Recording, Maintenance *and* Management *for the* Biomedical Sciences

Shayne C. Gad and
Stephanie M. Taulbee

CRC Press

Boca Raton   New York   London   Tokyo

Publisher:	Bob Stern
Assistant Managing Editor:	Gerry Jaffe
Marketing Manager:	Susie Carlisle
Direct Marketing Manager:	Becky McEldowney
Cover Design:	Denise Craig
Pre Press:	Kevin Luong
Manufacturing:	Sheri Schwartz

Library of Congress Cataloging-in-Publication Data

Gad, Shayne C., 1948–
    Handbook of data recording, maintenance and management for the
biomedical sciences / Shayne C. Gad and Stephanie M. Taulbee.
       p. cm.
    Includes bibliographical references and index.
    ISBN 0-8493--0137-8 (alk. paper)
    1. Medical sciences--Research--Documentation. 2. Life sciences-
-Research--Documention. 3. Fraud in science--Prevention.
I. Taulbee, Stephanie M. II. Title.
R852.G33   1996
610′.72--dc20
                          96-3960
                          CIP

9

Authors

Shayne C. Gad, Ph.D., DABT, is the Principal of Gad Consulting Services in Raleigh, North Carolina, a consulting firm providing services to the pharmaceutical and medical device industries in the area of safety assessment, regulatory considerations, statistics and risk assessment. His research interests include alternative testing and models, neurotoxicology, cardiovascular, dermal, and immune toxicology. He has published 15 books and more than 275 independent chapters, papers, and abstracts. He is Editor-in-Chief for *Toxicology Methods* and is on numerous editorial boards. He has served on the NIST Combustion Toxicology Task Force, the Consumer Product Safety Commission Toxicology Advisory Board, the NIH Occupational Safety and Health Study Section, the SOT Animals in Research, Nominations and Placement Committees, the American College of Toxicology (ACT) Council, Membership and Animals in Research Committees, and on trade association panels for nylon 6, chrome chemicals, cyclohexanone, ketones, phthalates, latex, and medical devices. He is a past president of the American College of Toxicology.

Dr. Gad has lectured at the University of Texas, the University of Kansas, Rutgers, Johns Hopkins University and the University of Pittsburgh, and is on the faculty of Duke University. He has served on several Ph.D. thesis committees and numerous grant and special review panels and boards for the EPA, DOE, NIEHS, Superfund, NIH, Johns Hopkins, CAAT, and the Canadian government. He has also established and taught a bachelor program in toxicology at the College of St. Elizabeth.

Stephanie M. Taulbee, M.S.P.H., is Manager of the Quality Assurance Unit for the Chemistry and Life Sciences Unit at Research Triangle Institute, Research Triangle Park, North Carolina.

Ms. Taulbee graduated in 1991 from North Carolina State University, Raleigh, with a B.S. in Health Occupations Education (summa cum laude) and obtained her M.S.P.H. in 1996 in Environmental Sciences and Engineering from the University of North Carolina at Chapel Hill.

Ms. Taulbee received Research Triangle Institute's Professional Development Award to complete her masters research, "A Critical Review of Published Data for Use in Developmental and Reproductive Toxicity Risk Assessments."

Ms. Taulbee is a member of the Society of Quality Assurance and the North Carolina Quality Assurance Discussion Group (NCQADG). Through these organizations she has presented seminars and posters at annual and local meetings. She served as Training Subcommittee Chair from 1991 to 1993 for the NCQADG. In 1994, she chaired and organized a seminar, The Importance of Scientific Accuracy: Issues of Ethics and Integrity, co-sponsored by NCQADG and North Carolina Chapter of Sigma Xi.

Ms. Taulbee has co-authored four books and several journal articles on the U.S. and International Good Laboratory Practices (GLPs). She has developed a training program on the GLPs that has been presented at Research Triangle Institute and at several professional meetings. Her current interests are in developing expanded resource and training materials on computer validation and quality assurance in Regulatory Toxicology.

Contents

chapter one

Introduction

Shayne C. Gad

All science, but particularly the biomedical sciences, has come under increased suspicion and scrutiny since the 1960s. This is due to a variety of reasons (particularly a more public airing of problems in a number of laboratories) and has necessitated all scientists taking better care (and having to meet some expected minimum standards) in how they document their work, results, and observations. A good starting place is to consider the philosophical principles under which science operates and how they impact on the need to keep records.

Basis of modern science

There are some fundamental operating principles which have led to the success and creditability of modern science and which are fundamental philosophical underpinnings for all of the scientific disciplines (Kryburg, 1968; Broad and Wade, 1982). These can be summarized in the six points below.

Replication: It must be possible to repeat any experiment or observation for it to be accepted into the fabric of scientific knowledge. In a formal sense, it is assumed that all experiments and observations are/will be replicated, though in the modern era this is clearly not the case. Replication clearly requires, however, that the documentation associated with scientific work be adequate so that (1) the procedures, methods, and materials utilized can be identified and duplicated and (2) that the original data or observations resulting from an experiment can be independently analyzed and conclusions verified. Having a second person review protocols and data recordings in a timely manner to make sure that they are complete and clear goes a long way towards ensuring that replication is possible at a later date.

Falsifiability: Theories, hypotheses, and conclusions in science must be of such a nature that it is possible to show them not to be true — that is, to show that they are false. It is not possible to prove a true negative — that is, to show that something absolutely is not the case. Experiments and observations must therefore be of such a nature that sensitivity and precision of measurements are sufficient to allow replication and described work to be the expected common outcome. And a conclusion must be such that it can be shown not to be true. A conclusion or theory is not established by the inability of others to show it not to be true when such a proof is either not possible or beyond the limits of current technology. An example in modern toxicology is the theory that there is no threshold for carcinogens — that is, no exposure below which they do not act to cause cancer. Inability to prove that this is not the case does not establish that it is.

Induction and experiment: Induction is the act of deriving our understanding of nature, the universe, and truth by direct observation and experience. Such induction is probabilistic in that it is assumed that a sufficient (objectively defined) amount of data or observations of a like sort allow us to have a (statistically defined) level of confidence that they are representative of the universe as a whole. Modern science is founded on the use of inductive reasoning and the conceptual tool used to provide sufficient objective evidence to use it. This conceptual tool is the experiment, of course. Such is the marriage of modern science and statistics. Our research must be conducted in such a manner (and records must reflect this) that statistical validity is achieved and maintained. More details on the statistical aspects of experimental design, conduct and analysis can be found elsewhere (Gad et al., 1988).

Law of parsimony: Also known as Ockham's razor, the formal statement is that entities are not to be multiplied beyond necessity. Conclusions and theories that are the most straightforward — which do not require extra elements, mechanisms, or events are to be preferred. Likewise, those procedures, systems, and forms that are simplest and the most straightforward are to be preferred for record keeping and documentation.

Correlation and causality: Correlation is the associated occurrence of two events or of an event and an observation. It can be statistically evaluated for significance and can be either positive ("when A is done, B happens") or negative ("when A is done, B doesn't happen"). Descriptive science depends heavily on correlation. But because two events are correlated does not establish that one occurs because of the other — that A causes B. Mechanistic science seeks to determine what causes things. A knowledge of correlation of events can provide powerful means to understand the mechanisms ("causes") underlying the events or observations. But only by establishing causes can we understand mechanisms, which tend to provide the foundation for broader understanding of and theories about the natural world.

History

Good Laboratory Practices (GLP) as law have been with us since 1977, with the primary need that they were intended to meet both preceding that date by many years and continuing to the present.

Good recording of data, plans, and procedures in the laboratory has always been essential to the conduct of both scientific research and the entire self modifying/evolutionary process by which science as a whole operates. The documentation of the fact that such procedures were followed is an unfortunate reflection of the need to insure against everything from sloppiness to dishonesty, and in many areas of biomedical research and testing, is now also a requirement of law. To understand the need both for all of these procedures and for the laws requiring them, we must review the history of problems in the area.

A complete history of the problems associated with biomedical data recording and management is a book in itself. In fact, a number of books have been published on this very matter (Broad and Wade, 1982; Huber, 1991). Though the problem of data falsification (or the suspicion of such) dates back to Ptolemy and is not limited to the biomedical sciences, our overview of history will be limited to from 1960 on and to the biomedical sciences.

In 1960–1961, a graduate student (at Yale) who went on to become a postdoc (at Rockefeller) performed brilliant experiments (the results of which were widely published) on cytochrome c and glutathione with well-respected senior investigators

(Broad and Wade, 1982). The work was soon found not to be repeatable, the articles were retracted, and the junior individual involved resigned and left research. The episode received no press attention.

The first widely publicized case to come to the public's attention (and to start to erode the publics faith in science) was that of the "patchwork mouse" in 1974 (Hixson, 1976). William T. Summerlin was a junior researcher working at Sloan-Kettering in a large lab run by Robert Good. Good's lab had published almost 700 well-regarded papers in immunology (with Good as a co-author on all of them) over the preceding 5 years. Summerlin reported a number of successful transplantations in animals which could not be replicated by others. Finally, he used a black felt tip pen to enhance the appearance of successful transplantation of skin patches on some mice. A technician (the laboratory supervisor, but not Dr. Good) detected the alteration in what became a well-publicized case.

In 1978 an entire team of researchers working for Dr. Marc Straus at Boston University were working as part of a clinical trial sponsored by the Eastern Cooperative Oncology Group. The team reported that they had "falsified" nearly 15% of all the data entered from the trial, under direction from Dr. Straus. The falsification consisted of everything ranging from concealing errors made by the team in following the specific study protocol to allowing physicians to diverge from the study treatment without having to exclude the patients from the trial. This situation was repeated with much wider publicity by a Canadian research team that was part of the breast cancer trials in 1994, leading to the well-respected overall head of the trial having to resign (Anderson, 1994).

Industry has had its share of problems, both real and suspected. During the 1970s the largest industry biological testing lab in the country was Industrial Biotest. In 1975 an FDA investigator stumbled by accident on problems in the data from testing on Naprosyn. As investigators dug deeper into the data on studies on the safety of more than 600 drugs, chemicals, and food additives evaluated by IBT, they found enough fraud to lead to the indictment and conviction of four senior officers of the company (Anon., 1981a,b; 1983a,c). Of greater impact was that the documentation of study procedures and data recording could not be verified. Given the known problems with the data and the conduct of some studies (animals that died several times, daily body weights recorded after animals had died, animals that died on study not being necropsied until after autolysis had set in, etc.), the results of all the studies were suspect. Studies either had to be repeated or (if possible) validated. This case and others in the same time frame led to the adoption of the Good Laboratory Practice (GLP) regulations which now govern all preclinical (i.e., nonhuman) studies performed to establish the safety of a drug, medical device, or chemical regulated by the U.S. and most foreign governments.

The GLP regulations, which are discussed in a later section of this book, call for regular inspections of all laboratories (industry, contract, and university) involved in the generation of such data. This regular program of inspection has continued to identify problems — some of actual fraud (inventing or altering data) — and others of violation of procedural/documentation requirements of the regulations. A few examples spanning the first 15 years since the regulation became effective are given in Table 1.1.

The problems which have led to a decrease in the creditability of science have not been limited to industry. As shown in Table 1.2, academic and government labs and researchers have also had problems on a continuing basis. These problems have not been just of fraud (real and suspected), but also of plagiarism and various other forms of scientific misconduct. Scientific misconduct has a variety of forms:

Table 1.1   Industry and Contract Lab Violations of the GLPs (1980–1994)

Organization	Year	Violation	Penalty
Litton	1980	Deviations from protocols and SOPs Mixup or misidentification of test materials Inadequate SOPs	Warning letter
Gulf South Research	1983	Poor data keeping on NTP carcinogenicity studies	Lab went out of business
Biodynamics	1980	Timeliness of postmortem exams Reporting of tumors Poor husbandry	None
	1983	Late reporting Pathologist not present at necropsy Poor husbandry	None
SAIC	1986	Backdating of Superfund data	$750,000 fine
Carter Wallace/AMA Laboratories	1992	No study protocols Failure to sign data entries No study personnel files	$132,000 fine
Bio-Tek Industries/Microbac Laboratories	1992	No QA unit Lack of written protocols and SOPs Missing items and inconsistencies in raw data and report	$100,000 fine
Craven Laboratories	1992	"Tweaking" of pesticide residue data	Prison terms
Twelve pesticide firms	1993	Inadequate documentation and records	$183,000 in fines

Table 1.2   Proported Recent Cases of Academic and Government Scientific Misconduct

Institution	Year	Allegations	Outcome
Tufts	1986	Fraudulent data in *Cell* paper	Secret Service involved Five years of investigations NIH finding of fraud
Vanderbilt	1982	Fraud, poor record keeping	Discrediting of research on alcoholism (and of researchers)
Caltech	1989	Fraudulent and missing data	Paper retracted Responsible postdocs dismissed
University of Pittsburgh	1979	Misanalysis of lead data	Office of Scientific Integrity investigation
University of California, San Diego	1986	Publishing false data	Faculty member resigned
University of Alabama, Birmingham	1989	Plagiarism False claims to the government	$2MM civil suit verdict
St. Luc Hospital/ Montreal	1994	Not following protocol Falsifying ineligible patient enrollment	Overall breast cancer study head removed

Plagiarism: Presenting work done by another as your own.

Misallocation of credit: Claiming (or accepting) credit for work done by another. This includes a lack of adequate acknowledgment of the work of one's intellectual predecessors.

Bias: Uneven, unbalanced, or onesided collection, analysis, or reporting of data.

Trimming: Improving the appearance of quality of work or of clarity of outcome by removing or failing to report some data or observations.

Sloppy/poor records and methods: The most honest of intentions, but the documentation of what has been done and seen is either so incomplete, unclear, or disorganized that the value of the work is at best suspect and discounted.

Wholesale fraud: Complete invention of some or all of the work done and resulting data.

Junk science: That which supports adversarial opinions and is not supported by the work of others or accepted by the scientific community (Huber, 1991).

Principles

There are some basic guidelines which, if not observed, contribute to the occurrence of data/documentation problems when they are not the result of intentional fraud. It is particularly difficult for many academic laboratories to adhere to all of these guidelines, as to do so requires that there be more control or oversight of laboratory operations (and therefore intrusion on "creativity" and "academic freedom"). But the results of failing to honor these guidelines are listed in Tables 1.1 and 1.2 with potential costs, as will be discussed later in this chapter.

1. Keep laboratories small enough to allow adequate oversight and supervision. Each person in a supervisory role (whether a group leader in industry or a principal investigator in an academic setting) has what is characteristically called a span of control — a number of individuals and range of operations that he or she can successfully monitor, provide leadership for, and regularly review to insure that all is as it should be. The "too large lab syndrome" occurs when these responsibilities are accepted which exceed the span of control. Most supervisors can effectively directly oversee the work of four to eight others. Some can effectively directly supervise as many as 13 to 15. Span is very much an individual characteristic. But no one can effectively supervise 150 graduate students and postdocs, as has sometimes been attempted.
2. Document what you do and see as you do it and see it. If it isn't on paper (or an appropriate electronic medium), it didn't happen. And as part of the documentation, record who wrote it down and when.
3. Verify documentation. For legal reasons and to provide a check against simple human lapses, always have a second individual verify that documentation exists, is adequate, and was created at the time the work was done or observations made.
4. Elimination/control of error. Independent of the actual daily process of research, someone needs to study the kinds of errors which are occurring and the factors which cause them. Errors happen for reasons, and as important as catching and correcting errors is making changes to eliminate their sources.

Costs of inadequate record keeping

Following the guidelines above (and the practices presented in the body of this volume for implementing these guidelines) clearly has a cost. Proper record keeping and documentation is not free — it takes time, effort, and an increment of material resources that could be invested elsewhere. Many experienced industry researchers who operate under GLPs estimate that doing so represents ~25% of the total workload. Good record keeping in a nonregulated environment should not be that expensive, but will require a significant investment. What justifies such a diversion of resources?

Beyond the basic tenet that current accepted scientific practice and standards of research conduct require that records and documentation meet the guidelines presented earlier (that is, that it is the right thing to do), there are some very real costs in not adhering to such standards. These include:

1. Loss of grant money or other continued funding.
2. Inability to patent intellectual property resulting from research (see King, 1995 for a recent example) .
3. Discredit before one's peers. For the most extreme recent examples, one need only have followed the case of David Baltimore's lab, where in the end a well-respected investigator suffered due to the actions of a subordinate (Hall, 1991; Hamilton, 1991; Maddox, 1991; Stone and Marshall, 1994) or Robert Gallo's work on HIV (Cohen, 1991; Palca, 1991a,b). Ultimately, as scientists the one greatest (perhaps only true) asset we have is our creditability.
4. Loss of position (as happened with Fisher having to resign as head of the breast cancer study due to the actions of others — Anderson, 1994).
5. Criminal (felony) charges (Zurer, 1991).
6. Civil suit by other researchers (Taubes, 1995).

If not for the sake of convictions, then to avoid the consequences above, the biomedical scientist must meet the expected standards of record keeping and documentation. The remainder of this volume is intended to provide practical guidance on how to do so.

References

Anderson, C. (1994), How not to publicize a misconduct finding, *Science* 263:1679.

Anon. (1980), FDA tells Litton to bring three facilities into GLP compliance, *Food Chem. News,* May 12, pp.19–20.

Anon. (1981a), Nalco Chemical unit ex-officials charged with faking lab data, *Wall Street J.,* June 23.

Anon. (1981b), Lab execs indicted for faking toxicity data, *Chem. Eng. News,* June 29, p. 5.

Anon. (1983a), The darker side of a laboratory, *Chem. Week,* May 18.

Anon. (1983b), Bio/Dynamics defends permethrin studies; House subcommittee may investigate, *Pestic. Toxic Chem. News,* September 21, pp. 31–32.

Anon. (1983c), Tighter controls on toxics testing, *Chem. Week,* August 24, pp. 32–39.

Anon. (1983d), There are no plaudits for EPA laboratory audits, *Ind. Chem. News,* June, pp. 30–33.

Anon. (1991), NIH finds misconduct at Georgetown, *Science* 252:35.

Anon. (1994), Twelve companies violated FIFRA Good Lab Standards, EPA says, *Pestic. Toxic Chem. News,* October 5, pp. 13–14.

Broad, W. and Wade, N. (1982), *Betrayers of the Truth*, Simon & Schuster, New York.

Carson, P. A. and Dent, N. J. (1990), *Good Laboratory and Clinical Practices*, Heinemann Newnes, Oxford.

Cohen, J. (1991), What next in the Gallo Case?, *Science* 254:944–949.

Cross-Smiecinski, A. and Stetzenbach, L. (1994), *Quality Planning for the Life Science Researcher*, CRC Press, Boca Raton, FL.

Gad, S. C. and Weil, C. S. (1988), *Statistics and Experimental Design for Toxicologists,* Telford Press, Caldwell, NJ.

Garfield, E. (1988), The international school of professional ethics: or, How to succeed in science without really trying, *Curr. Contents,* February 22, pp. 3–5.

Hall, S. S. (1991), Baltimore resigns at Rockefeller, *Science* 254:1447.

Hamilton, D. P. (1991), Verdict in sight in the "Baltimore Case," *Science* 251:1168–1172.

Hixson, J. (1976), *The Patchwork Mouse,* Doubleday, New York.

Huber, P. W. (1991), *Galileo's Revenge,* Harper Collins, New York, 274 pp.

King, R. T. (1995), Expert calls Calgene research on gene-altering method flawed, *Wall Street J.,* April 24, p. B4.

Kumar, V. (1991), Hood Lab investigation, *Science* 254:1090–1091.

Kyburg, H. E. (1968), *Philosophy of Science,* Macmillan, New York.

Maddox, J. (1991), Secret Service as ultimate referee, *Nature* 350:553.

Marcus, F. F. (1983), U.S. says sloppy drug tests at laboratory in south prompt wide review, *Wall Street J.,* October 24, p. D12.

Palca, J. (1991a), The true source of HIV?, *Science* 252:771.

Palca, J. (1991b), Draft of Gallo report sees the light of day, *Science* 253:1347–1348.

Pendery, M. L., Maltzman, I. M., and West, L. J. (1982), Controlled drinking by alcoholics? New findings and a reevaluation of a major affirmative study, *Science* 217:169–175.

Roberts, L. (1991), Misconduct: Caltech's trial by fire, *Science* 253:1344–1347.

Sobell, M. B. and Sobell, L. C. (1978), *Behavioral Treatment of Alcohol Problems,* Plenum Press, New York, p. 225.

Stone, R. (1991), Court test for plagiarism detector?, *Science* 254:1448.

Stone, R. and Marshall, E. (1994), Imanishi-Kari case: ORI finds fraud, *Science* 266:1468–1469.

Taubes, G. (1995), Plagiarism suit wins; experts hope it won't set a trend, *Science* 268:1125.

Tifft, S. (1991), Scandal in the laboratories, *Time,* March 18, pp. 74–75.

Tokay, B. A. (1984), Keeping tabs on the toxicology labs, *Chem. Bus.,* February, pp. 12–20.

Zurer, P. S. (1991), Contract labs charged with fraud in analysis of Superfund samples, *Chem. Eng. News,* February 25, pp. 14–16.

chapter two

Data recording

Stephanie M. Taulbee

"Good science" is a term used in much the same way as "good music." It is used to describe competent and reliable scientific conduct, although sometimes the use has as varied a meaning and interpretation as does "good music." It refers to many facets of scientific inquiry — study design, methods, data collection, reporting of results, and extrapolation of the study conclusions. You will hear that "it's only good science to ...." But what constitutes "good science?" We chose in this book to focus on the "good science" of data collection because, if all else is flawed, the data *if properly collected* are available and may be reevaluated. *Ipsa loquitur* — the thing speaks for itself. That is, the data will tell the story if researchers take seriously their responsibility to preserve the observations made during scientific inquiry.

Stories in scientific history attest to the necessity of maintaining thorough records of experimental results. Whether an experiment suceeds or fails, valuable insights may be gained. The preservation of this information allows the researcher to evaluate and improve methods in subsequent work and, on those wonderful, rare moments of discovery, to stumble upon a new and unique understanding. How often have important discoveries or precious research findings been lost because the experimenter did not record an observation, a step in a procedure, an assumption that, when reassessed, may have led to a new understanding!

This text is designed to sensitize the student of science to data collection, data maintenance, and to data quality. In this chapter, we discuss how data are used and developed, types of data, and methods for writing and recording data.

Data

What are data? Webster's dictionary defines data as "any fact assumed to be a matter of direct observation" or "any proposition assumed or given from which conclusions may be drawn" (Webster, 1983). For our purposes, data include records of methods, procedures, observations, calculations, and conclusions. The written record of data allows information to be processed, analyzed, and communicated, thereby contributing to the body of scientific knowledge.

How data are used

For now, let us consider data in biomedical research. Data are used to support scientific conclusions which may be used to increase understanding of some theory, describe a phenomenon or mechanism, or dispute a previously held understanding.

These are also used to support requests for additional research funds. Data are used to justify patent applications and claims, and are reported to government agencies to satisfy regulatory requirements. They are presented at scientific meetings, published in journals, described in books. Beyond this, data are reviewed and evaluated by other researchers in the field and are sometimes combined with other data to form broader and more far-reaching conclusions. Clearly, then, the uses of data dictate that they be accurate, and therefore, reliable.

These points may seem excruciatingly obvious but are of such great importance that it is necessary to remind ourselves of them. Human memory has been well documented to be unreliable. The written record of research activities and observation is designed to preserve data both for the researchers' intended use and for posterity. The scientist must record every aspect of research to preserve it.

However data are used, there will be requirements for reporting the data and conclusions: journals require a certain format for describing methods, results, and conclusions; government agencies publish guidelines for specific testing and study designs and have legislated requirements for data collection, analysis, and report formats. The researcher must obtain these requirements, develop a thorough understanding of them, and follow them. For many types of biomedical research, potential legal liabilities must be considered in developing and reporting data. Figure 2.1 illustrates just some of the federal agencies that promulgate regulations, issue research guidance documents, and control research funding.

Department of Health and Human Services
Public Health Service
• Food and Drug Administration
  - National Center for Toxicological Research
  - Center for Biologics Evaluation and Research
  - Center for Food Safety and Applied Nutrition
• National Institutes of Health
  - National Cancer Institute
  - National Institute for Environmental Health Sciences
• Centers for Disease Control
  - Agency for Toxic Substances and Disease Registry

Occupational Safety and Health Administration

Department of Agriculture

National Science Foundation
Office of Science and Technology

Department of Justice
Drug Enforcement Administration

Department of Defense
Department of the Army
Department of the Navy

Environmental Protection Agency
Air and Radiation
Water
Solid Waste and Emergency Response
Prevention Pesticides and Toxic Substances
Office of Research and Development

Figure 2.1 Some of the major sources of federal regulation and funding for biomedical research.

For the remainder of this chapter, the data standards discussed will be the Good Laboratory Practice Standards (GLPs) of the Food and Drug Administration (FDA) and the Environmental Protection Agency (EPA). The GLPs are federal regulations for development and reporting of data from safety assessments used to support marketing permits for drugs, food additives, pesticides, and toxic substances. While these standards are not universally required for all research efforts and specifically not required for development of research methods, they represent a sensible standard that may be applied to all data development practices (Figure 2.2).

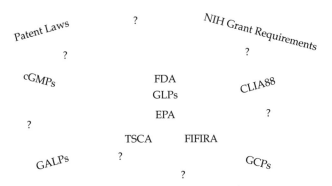

Figure 2.2 The complex world of regulated research.

Development of the study record

The objective of research is to evaluate theories and produce results. The written record of this work is called the study record and includes all records, documentation, and results of the research. Let us now consider the logical progression of research activities and the development of the study record.

Research plan

The research project begins with developing the research plan, or simply thinking through what will be done. Whether the worker is performing independent research, grant work, or research in support of regulatory requirements, this plan should be written down. When written, the research plan becomes the protocol for the project and includes the hypothesis, the proposed methods, observations to be made, and the expected results. Researchers should pay special attention to the level of detail in this plan. For example, in certain research environments there are requirements for inclusion of particular details in the protocol and a specified format. Optional experimental methods may be included in the protocol or amended into it as needed, but (again) must be recorded. Even if a written protocol is not specifically required for your research project, it is useful to develop the habit of producing a protocol because it requires you and your colleagues to think clearly through the experimental design. It also provides guidance for the actual conduct of the work and promotes consistency in performance (Figure 2.3).

Standard operating procedures

Some of the procedures performed during the study are routine for the laboratory. Formalize the documentation of these routine procedures into written standard operating procedures (SOPs). SOPs are detailed descriptions of such things as equip-

---

Research Plan or Protocol

Standard Operation Procedures (SOPs)

Recording Observations – Data Generation

Evaluation of Data

Report of Data, Results and Conclusion

Figure 2.3   Progression of a research project.

ment operation, methods for taking and recording data, and procedures for reagent receipt, storage, and preparation — the types of procedures that are common to all laboratory operations. Write SOPs in sufficient detail to promote consistency in performing the procedures. Having SOPs and insisting that they are followed provides the researcher with a measure of control over potential variables in the experiment.

Data recording

The experiment begins. You perform a procedure, write down what you did, and record the observed results. The level of detail of this written record should enable someone else with equivalent technical training to perform your experiment exactly as you did. Why? *Reproducibility.* That experimental results must be reproducible is a basic rule of science. It is the process through which scientific conclusions and discoveries are confirmed. Reproducibility is promoted by the specific data-recording requirements for data that are submitted to FDA, EPA, and other government agencies. Reproducibility is required in research performed to support a patent request.

For now, I wish to introduce you to the concept of "if you didn't write it down you didn't do it." You, the researcher, have the burden of proof in regulated research, in protection of patent rights, and in defense of your work in professional circles. The issue is *completeness* of your records. The study record must be a complete record of all data and procedures performed. If you didn't write it down, you didn't do it. In the experimental record, there are some accepted shortcuts. Here some of the hard preparatory work pays off. In your written record, you may include references to previously described methods and SOPs, state that they were followed exactly, or describe deviations from them. And, efficient ways of collecting data may be developed to encourage the complete recording of all required data. Later in this chapter, methods for recording procedures and observations will be discussed in detail.

The *accuracy* of recorded data is another important consideration because any observed result, if not recorded immediately, may not be recorded accurately. Don't lose data because of some rationalization about time, money, or your ability to remember what happened. All data should be recorded directly into a notebook or onto a worksheet at the time of the observation. Also, transcribed data — data copied by hand or entered by a person into a computer — often is subject to errors. If data are copied to a table or a spreadsheet, the entered data should be checked against the original data to ensure accuracy.

Analysis of the data

When the laboratory work is done, the researcher's analysis of the data begins. Observed data are entered into formulas, calculations are made, and statistical analysis is performed. All these manipulations must be carefully recorded, for from these data the conclusions will be drawn. The manipulations of the data are the link between the original observation and the conclusions. Consistency between the data and the result is controlled by monitoring all transcription, manipulation, and correlations of the data in generation of the final manuscript.

Reporting of results and conclusions

Finally, the manuscript or draft final report is submitted for peer review, to the publisher, or to the client. It will receive critical review before publication. The final version will then be published and again will receive critical review by your peers or some skeptical governmental or public audience. In all cases, it will be essential to be able to justify the data. The methods, initial data, the calculations and statistical analysis, and the conclusions must be defensible, meaning complete, accurate, internally consistent, and repeatable to withstand scientific criticism.

Types of data

Earlier, I mentioned different elements of the study record: research plan or protocol, observations, calculations and statistical output, and conclusions. For ease of explanation, the terminology from the GLPs — protocol, raw data, statistical analysis, and final report — will be used to describe the components of the study record.

From the GLPs, the protocol is a written document that is approved by the study director (person responsible for the technical conduct of the study) and sponsoring organization. The protocol is the research plan, the approved grant proposal, the project plan in a management sense. It clearly indicates the objectives of the research project and describes all methods for the conduct of the work. It includes a complete description of the test system, the test article, the experimental design, the data to be collected, the type and frequency of tests, and planned statistical analysis.

The protocol will be strictly followed during research. "What," you say, "no experimental license, no free expression of scientific inquiry?" Of course there is, as long as the changes in procedures or methods are documented. If the work you are doing is governed by strict contractual or regulatory guidelines, you may not be able to express much creativity, but remember, the objective, in this case, is to provide consistent reliable comparisons for regulatory purposes. Even the GLPs make provisions to amend the protocol and document deviations from it. During all research, except perhaps during the most routine analysis, there may be changes in experimental methods and procedures, rethinking of design, decisions to analyze data in new or different ways, or unexpected occurrences that cause mistakes to be made. An important concept to apply here is that these variances from the plan must be documented.

Raw data

"Raw data" is the term used to describe the most basic element of experimental observations. It is important to understand fully the concept of raw data. There are unique standards for recording raw data that do not apply to other types of data. These will be discussed later in the chapter. For now, let us look at what constitutes raw data. In the FDA and EPA GLPs,

> … raw data means any laboratory worksheets, records, memoranda, notes or exact copies thereof that are the result of original observation and activities of the study and are necessary for the reconstruction and evaluation of the report of that study.

All terms must be taken in the most literal sense and must be interpreted collectively to apply this definition to the data generated during an experiment. There are two key phrases: "are the result of original observations and activities of the study," and "are necessary for the reconstruction and evaluation of the report of that study." Raw data include visual observations, measurements, output of instrumental measurements, and any activity that describes or has an impact on the observations. Anything that is produced or observed during the study that is necessary to reconstruct (know what happened) and evaluate (analyze or, for regulatory purposes, assess the quality of) the reported results and conclusions is raw data. This definition of raw data has been carefully designed to encourage the development of data that are defensible.

Included in the scope of raw data may be data that result from calculations that allow the data to be analyzed, for example, the results of gas chromatography where the raw data are defined as the curve that was fitted by the instrument software from individual points. The individual points on the curve are essentially meaningless by themselves, but the curve provides the needed basic information. The area under the curve, which is used to calculate the concentration, is an interpretation of the curve based on decisions made about the position of the baseline and the height of the peak. This is not "raw data" since it is not the original observation and may be calculated later and, practically, may be recalculated. For the researcher to completely understand the results, the curve with the baseline, the area under the curve, and the calculations are required and recorded, but only the curve itself is "raw data." The distinction is that the curve is the original observation and must be recorded promptly.

Other types of data

Other types of data that are not thought of as raw data may be included here. For example, correspondence, memoranda, and notes may include information that is necessary to reconstruct and evaluate the reported results and conclusions. While these are not records of original experimental observations, they do represent documentation of the activities of the study. They often contain approvals for method changes by study management or sponsoring organizations, instructions to laboratory staff for performing procedures, or ideas recorded during the work. Here are some examples of raw data that are generated during a toxicology study:

Test article receipt documents	Equipment use and calibration
Animal receipt documents	Equipment maintenance
Records of quarantine	Transfer of sample custody
Dose formulation records	Sample randomization
Sample collection records	Animal or sample identification
Dosing records	Assignment to study
Animal observations	Necropsy records
Blood collection and analysis	Analytical results
Euthanasia records	Histology records
Pathologist's findings	

For government-regulated research, all records that are documentation of the study conduct are treated as raw data. From the perspective of the scientific historian, the original notes, correspondence, and observations tell the story of the life and thought processes of the scientist being studied. From the mundane to the extreme, these records are important.

Computerized data collection

Special attention must be dedicated to computer-generated raw data. Automated laboratory instrumentation has come into widespread use. In hand-recorded data, the record of the original observation is raw data. But what is considered raw data in computerized systems? In this case, raw data are the first recorded occurrence of the original observation that is human readable. This definition treats computer-generated data as hand-recorded data. It documents the "original observations and activities of the study and is necessary for reconstruction and evaluation of the report of that study" (FDA, 1987; EPA, 1989). However, we must pay special attention to this type of data. The validity of hand-recorded data is based on the reliability of the observer and on well-developed and validated standards of measurement. For computer-generated data, the observer is a computerized data collection system, and the measurements are controlled by a computer program. These are complex systems that may contain complex flaws. Just as the principles behind measurements with a standard thermometer were validated centuries ago and are verified with each thermometer produced today, so must modern computerized instrumentation be validated and its operation verified. This causes a real dilemma for many scientists who are proficient in biomedical research but not in computer science. Because of the size and scope of this issue, I can only call your attention to the problem and refer you to the literature for additional guidance. I will discuss special issues in recording raw data, including computer-generated raw data, later in this chapter.

Statistical data

Statistical data result from descriptive processes, summarization of raw data, and statistical analysis. Simply put, these data are not raw data but represent manipulation of the data. However, during this analysis process, a number of situations may effect the raw data and the final conclusions. For example, certain data may be rejected because they are shown to be experimentally flawed, an outlier believed to have resulted from an error, or not plausible. I will leave it to other texts to discuss the criteria by which decisions like these are made. Here, I will say only that any manipulating of raw data is itself raw data. For example, a series of organ weights is analyzed. One of the weights is clearly out of the usual range for the species, and no necropsy observations indicated the organ was of unusual size. The preserved tissues are checked, and the organ appears to be the same size as others in the group. The statistician then may decide to remove that organ weight from the set of weights. This record of this action is raw data. The analysis is not, because it can be replicated. It is a fine distinction that matters only in the context of recording requirements for raw data since both the analyses and record of the data change are required to reconstruct the report.

Statistical analysis is part of the study record. Documentation of the methods of statistical analysis, statistical parameters, and calculations is important. Critical evaluation of conclusions often involves discussion of the statistical methods employed. Complete documentation and reporting of these methods, calculations, and results allows for constructive, useful critical review.

Results and conclusions

The study record includes the results and conclusions made from review of the data produced during the scientific investigation. The data are summarized in abstracts, presented at meetings, published in journals, and, with all previously discussed types of data, are reported to government agencies. However, it is the scientist's interpretation of the data that communicates the significance of the experimentation. In all scientific forums, scientists present their interpretation of the data as results and conclusions. Results and conclusions are separate concepts. This is an important distinction not only because it is the required format for journal articles and reports, but because it is important to separate them in one's understanding. Results are a literal, objective description of the observations made during the study, a statement of the facts. Conclusions, on the other hand, represent the analysis of the significance of these observations. They state the researcher's interpretion of the results. If results are presented clearly and objectively, they can be analyzed by any knowledgeable scientist, thereby testing the conclusions drawn. This is the process by which the body of scientific knowledge is refined and perfected.

For regulatory purposes, the results presented to the regulatory agencies (FDA or EPA) must be complete. Included in the reports submitted are tables of raw data, all factors that affect the data, and summaries of the data. In journals, the results section usually is a discussion with tabular or graphical presentations of what the researcher considers relevant data to support the conclusions. Conclusions presented in either case interpret the data, discuss the significance of the data, and describe the rationale for reaching the stated conclusions. In both cases, the results are reviewed and the conclusions analyzed by scientific peers. The function of the peer review process is to question and dispute or confirm the information gained from the experiment. Objective reporting of results and clear discussion of conclusions are required to successfully communicate the scientist's perspective to the scientific community.

Development of study data

Above we have discussed the types of data that make up the study record. The following discussion addresses: quality characteristics for the study record, requirements for recording raw data, and methods for fulfilling the quality characteristics and raw data requirements by using various recordkeeping formats.

Quality characteristics

There are four characteristics the study record must have: completeness, consistency, accuracy, and reconstructability. *Completeness* means the information is totally there, self-explanatory, and whole. *Consistency* in the study record means that there is "reasonable agreement between different records containing the same information" (DeWoskin, 1995). *Accuracy* is agreement between what is observed and what is recorded. The final characteristic is *reconstructability*. Can the data record guide the researcher or someone else through the events of the study? These characteristics are goals to meet in developing the study record and will be used in chapter four to evaluate the quality of these records. They must be built into the study from the beginning, and considerable attention to these goals will be required as the study progresses to produce a complete, consistent, accurate, and reconstructable study record.

Recording raw data

Raw data may be recorded by hand in laboratory notebooks and worksheets or entered into a computerized data management system. Today, more and more data are computer generated and recorded as paper outputs or are electronically written to magnetic media, microfiche, or other storage media. This section will discuss how raw data in both forms are recorded.

General requirements for raw data recording

Raw data must be recorded properly to preserve and protect them. The following is excerpted from the FDA GLPs:

> All data generated during the conduct of a study, except those that are generated by automated data collection systems, shall be recorded *directly, promptly,* and *legibly in ink.* All data entries shall be *dated on the date of entry* and *signed or initialed by the person entering the data.* [Emphasis added.]

All introductory laboratory courses teach these basic techniques for recording raw data. Even though these standards are published as regulations for only certain types of research, I believe that there is never an instance when these minimum standards do not apply. There may be researchers who "get by" writing in pencil or scribbling data on paper towels, but they often suffer the consequences of their carelessness when data are lost or their records are unintelligible. Too, if these same researchers attempt to patent a product or method, or to submit their data to regulatory agencies, their data are not acceptable. In fact, if the regulatory data are incomplete or obscured in some way, the scientist involved may be subject to civil or criminal penalties. Chapter one discusses many related instances. It is always best to establish good habits early, especially for scientific recordkeeping.

For hand-recorded data, "directly, promptly, and legibly in ink" means to write it down in the notebook or on the worksheet as soon as you see it, so it is readable and in ink. The purpose is to preserve accurately the observation. Notes on paper towels or scratch paper may be lost. Prompt recording promotes accuracy. Legibility assures that later you will understand what is written. This does not necessarily mean neat. If you are recording directly and promptly, neatness may have to be forgone. It does, however, mean readable and understandable.

The use of ink preserves the record from being erased or smeared. It is commonly understood that the ink should be indelible, meaning it cannot be erased and can withstand water or solvent spills. Some organizations may require a specific color of ink to be used, usually black or dark blue. This requirement originated because black ink was the most permanent and could be photocopied. Without such requirements, the ink used in the lab should be tested to see how it withstands common spills and to see if it copies on the standard photocopier. Some colors of ink and some thin line pens may not copy completely. There are a number of reasons why data may need to be copied, and that they are copied exactly becomes a very practical issue. Inks should not fade with time. Some analytical instruments produce printed data on heat sensitive paper. To preserve these data, laboratories will make photocopies. This is an issue that will be discussed more fully in chapter three.

The requirements to sign and date the data record flow from practical and legal considerations; it is often useful to know who made and recorded the observation. In many research labs, graduate assistants or research technicians are responsible for recording the raw data. If questions arise later, the individual responsible may be sought out and asked to clarify an entry. For GLP studies, the signature represents a legal declaration meaning the data recorded here are correct and complete. The data must be dated at the time of entry. This attests to the date of the recording of the observation and the progression in time of the study conduct. Some lab work is time dependent and in this case the time and date must be recorded. There is no instance when data or signatures may be backdated or dated in advance.

Signatures and dates are crucial when documenting discovery and in supporting a patent claim. For studies conducted under the GLPs, the signature and date are legal requirements for the reconstruction of the study conduct. Falsely reported data may then result in civil or criminal penalties to the person recording the data and his/her management for making false and misleading statements.

In some types of research, additional signatures and dates may be required. Data used to support a patent and data generated during the manufacture of drugs or medical devices must be signed and dated by an additional person — a witness or reviewer thus corroborating the stated information.

Error correction in data recording

What happens when there is a mistake in recording data or an addition that must be made to the data at a later time? The FDA and EPA GLPs address this.

> Any changes to entries shall be made so as not to obscure the original entry, shall indicate the reason for such change, and shall be dated and signed at the time of the change.

All changes to the written record of the data must be explained and signed and dated. Doing so provides justification for the correction and again provides testimony as to who made the change and when it was done. To make corrections to the data, the original entry is not obscured. A single line is drawn through the entry. Then, the reason for the change is recorded with the date the change is made and the initials of the person making the change. A code may be established and documented to explain common reasons for making corrections to data. A simple example may be a circled letter designation like:

> S = sentence error
> E = entry error
> X = calculation error.

This is easy to remember and use. Any other types of errors or corrections must be described in sufficient detail to justify the change.

Raw data may be generated by computer programs and stored on paper or magnetic media. Most laboratories approach this kind of data as they would hand-generated data. The GLPs state:

> In automated data collection systems, the individual responsible for direct data input shall be identified at the time of the data input. Any changes to automated data entries shall be made so as not to obscure the original entry, shall indicate the reason for the change, shall be dated and the responsible individual shall be identified.

For automated data collection systems, there are similar standards to hand recorded data (FDA, 1987; EPA, 1989). All raw data should be recorded promptly and directly. Whereas the requirement for hand-collected data is that records be written legibly and in ink, permanence and security of computer-collected data is the requirement. However, there may be special considerations for how signatures and dates are recorded. Physical signature of data may not be possible when using electronic storage media. Electronic signature or the recording of the operator's name and the date are often a function provided in the software and are recorded with the data. When the data are printed in a paper copy, this information should be included. Some labs have adopted a policy requiring that the paper printout be signed and dated by the operator. Some instruments produce a continuous printout or strip chart. In this case, the chart should be signed by the operator and dated on the date the data are retrieved. If the data are maintained on electronic media, the operator's name and date must be recorded on that medium.

Because computer security and risk of corruption or destruction of computer-stored data are a major concern, many laboratories maintain computer-generated data in paper printouts because the means for maintaining the data are traditional and easy to implement. As long as the printout represents a verified exact copy of the original raw data, it is acceptable and often even preferable to designate the printout as the raw data.

When changes to the electronically stored raw data are made, the original observation must be maintained. This is accomplished in several ways. Newer software packages allow these changes to be made and properly documented. To do this, the original entry is not erased, and there is a way of recording the reason for the change along with the electronic signature of the person authorized to make the change and the date of the change. However, some data collection systems still do not have this capability. If this is the case, the original printout may be retained with the new printout that contains the change, the reason for the change, the signature of the person authorized to make the change, and the date of the change. Some computer programs allow for footnotes and addenda to be added to the record. These additions to the record, if made later, should also include a handwritten or computer-recorded signature and date.

Formats for recording data

We will now begin to construct the study record. The format for the study record may be determined by the preferences of the researcher. Some researchers prefer to maintain all study records in laboratory notebooks. In private industry, research and development labs may be required to use lab notebooks because of potential patent documentation requirements. Many chemists have become accustomed to the use of lab notebooks. However, handwritten data may be maintained in laboratory notebooks, on worksheets and forms, or one may use computer-generated printouts and electronic storage media. The remainder of this section discusses guidelines for recording data using all formats.

Laboratory notebooks

Laboratory notebooks are usually bound books with ruled or gridded pages that are used to record the events of an experiment. Organizations may order specially prepared notebooks that are uniquely numbered on the cover and spine. They have consecutively numbered pages, and some come with additional carbonless pages to make exact copies of the entries. Organizations may have procedures in place for issuing notebooks to individuals for use on specific research projects. After the glassware is cleaned, all that remains of a study is the notebook; its value is the cost of repeating all the work. Therefore, SOPs should be written to control the assignment, use, and location of these records.

The pages may be designed to contain formats for recording information. In the header, there may be space for the title and date. In the footer, space may be allocated for signature and date of the recorder, and signature and date of a reviewer or witness. When beginning to use a laboratory notebook, set aside the first few pages for the table of contents. Then a few pages may be held in reserve for notes, explanations, and definitions that are generally applicable to the contents.

The remainder of this section discusses the rules for recording data in the notebook. First, each page should contain a descriptive title of the experiment that includes the study designation and the experimental procedure to be performed. The date the procedure was performed is also recorded. Often a complete description of the experiment will require several pages. After the first page, subsequent pages should indicate, at least, an abbreviated title and cross reference to the page from which it was continued.

The body of the experimental record should include the following sections:

- Purpose of the experiment
- Materials needed, including instruments, equipment, and reagents
- Reagent and sample preparation
- Methods and procedures
- Results

The *purpose* may be recorded in a few sentences. The *materials section* is a list of all the things you need for the experiment — the instruments to be used, the equipment, and chemicals. When recording the analytical instruments, include the make, model number, and serial number; the location of the instrument; and all settings and conditions for the use of the instrument. The description of the chemical used should include a complete description including name, manufacturer, lot or serial number, and concentration. *Reagent and solution* preparation must be described in detail with a record of all weights and measurements. It is extremely important that sample identification and sample preparation be completely documented. The *methods and procedures* section is a step-by-step description of the conduct of the experiment.

If SOPs are in place that describe any of the above information in sufficient detail, they may be referenced. Then information recorded in the notebook is all weights and measurements, and any information that is unique to this experiment or not specifically discussed in the SOP. SOPs often are written for more general applications. An SOP may state that the pH will be adjusted using a buffer or acid as required. The notebook should indicate what was used to adjust the pH and how much was used. An SOP may describe the formulation of a compound in a certain amount, when the experiment requires a different amount. The mixing procedures may be cross-referenced, but it will be necessary to describe in detail the conversion of the SOP quantities and any changes in procedure resulting from the change in quantity.

The *experimental results* section must contain all observations and any information relating to those results. It should include any deviation from established methods, from SOPs, and from the protocol. Failed experiments must be reported even though the procedure was successfully repeated. Justification for repeating the procedure and a description of what may have gone wrong is recorded. All calculations should include a description of the formula used.

Remember, all entries are recorded directly and promptly into the notebook at the time of the experiment and are recorded legibly in ink. Some information may be entered at the beginning of the day, some entered at the end of the day, but all weights, measurements, and recorded observations must be entered into the notebook directly and promptly.

For a complete record, it is often necessary to insert such information as shipping receipts, photographs, and printouts into the lab notebook. In doing so, do not obscure any writing on the page. The following are tips for inserting information into the notebook.

- Glue the loose paper in place. (I do not recommend using tape because tape over time loses its holding power.)
- Inserts may be signed, dated, and cross-referenced to the notebook and page so that they can be replaced if they become loose.
- Make verified copies of data that is too large for the page, shrinking it to fit the notebook page.
- If, by some chance, data are accidentally recorded on a paper towel or other handy scrap of paper, these should be signed, dated, and glued into the notebook. It is not wise to transcribe data, introducing the possibility of error and the distasteful possibility of data tampering.

The bottom of each page must be signed by the person entering the data and dated at the time of entry. The date at the top of the page — the date of the activity — in most cases will be the same as the date at the bottom of the page. A few exceptions are appropriate. The most legitimate exception to this rule occurs when a page is reserved for the results printout. The printout may not be avaliable to insert until the following day. The printout should indicate the date when the data were first recorded, which should in turn match the top date. The date at the bottom of the page indicates when it was glued into the notebook.

Occasionally, a scientist will forget to sign and date the page. When this happens, there is no quick fix. The only remedy is to add a notation: "This page was not signed and dated on _____ , the time of entry," Then, sign and date this statement.

This discussion has been detailed because the signature and dates on the pages are very important. They are legally required for regulatory purposes. Data used to support patents and some data produced under the FDA current Good Manufacturing Practices (cGMPs) require the signature and date of a witness or reviewer. For example, the cGMPs require that all materials weighed or measured in the preparation of the drug be witnessed, signed, and dated. Patent applications are supported by witnessed experimental records. Some institutions may require supervisory review of notebook entries with accompanying signature and date. This is to say that you should be aware of the uses of your data and any requirements for this additional signature.

An important concept to remember is that bound, consecutively page-numbered notebooks are used to demonstrate the progression of the research and to document the dates of data entry and when the work was performed. To prevent the corruption of this record, unused and partially used pages may be marked out so no additions may be made. A suggested method is to draw a "Z" through the page or portion of the page not used. At the end of the project, there may be used notebook pages. These may be "Z'd," or the last page may indicate that this is the end of the experimental record and no additional pages will be used.

Forms and worksheets

While many analytical laboratories continue to use lab notebooks, other labs may use forms and worksheets to record their data. The purpose is to provide an efficient format for recording data that are routine in nature. Appendix one contains guidelines for the preparation of forms. The basic concept is that forms and worksheets should be designed to be easy to use and to provide a complete record of all relevant data. They may be used in combination with lab notebooks as described above or kept in files or loose-leaf binders. Explanatory footnotes may be preprinted or added to explain abbreviations and/or the meaning of symbols. Additional space for comments and notes should be incorporated into the format.

Computer spreadsheets and word processing make forms and worksheets easy to design and produce.

The advantages of using forms and worksheets include the following:

- They may be formatted to prompt for all necessary information.
- They are easy to follow and complete.
- Header information, title, study designation, sample numbers, etc. may be filled out in advance, thus saving time.
- Cross-references to applicable SOPs may be included on the worksheet.
- They help to standardize data collection.

Disadvantages of using forms and worksheets include the following:

- They must be carefully designed and should be pretested for completeness and ease of use.
- They may encourage a tendency not to write more information than is specifically requested. Space should be allotted for notes and comments.
- Forms and worksheets that are designed for general use may contain blanks that are not necessary for the current study. Yet all blanks must be completed. If not needed, "n/a" (not applicable) should be written in the blank or a dash put in the space.
- Forms and worksheets create a routine that can become mindless; take care to properly complete the form.
  - **Example 1**: Necropsy forms often contain a complete list of tissues to be checked by the technician. When only some tissues are inspected or retrieved, it may be too easy to check inappropriate boxes.
  - **Example 2**: Animal behavioral observation forms contain blanks to record all observations. The observer must record something in the blank space. A check or "OK" may be used for normal behavior if defined on the form or in an SOP. A problem occurs when these designations are used automatically without proper attention to observing and recording the behavior of each animal, particularly when most animals are behaving normally.

In discussing the above disadvantages, I'm not trying to discourage the use of worksheets. However, institute procedures and practices that assure that forms and worksheets are properly used.

As in any data record, the signature and the date of the entry are recorded at the time of the entry and represent and attest to the accuracy of the information. Any changes to the data or additional notes made after completion of the form or worksheet are made as previously described. Any unused lines on the form or worksheet should be crossed or "Z'd" out. If the signature of a witness or reviewer is required, there should be a line allocated for this purpose.

Forms and worksheets can be a useful and practical way to record and preserve raw data — if you pay attention to the rules of data recording.

Automated data collection systems

This is the hottest and most difficult topic of this book. Application of data collection rules to computer systems has been the topic of seminars, books, journal articles, government policy committees, and regulatory interpretation. As an example of the policy difficulties, the FDA has spent the last several years trying to reach consensus on a policy for electronic signatures. This policy may be published in the Federal Register by the time this book is published.

Two major issues surround automated data collection systems: validation of the system and verification of the system's proper operation.

Validation asks whether the system is properly designed and tested so that it performs as it should to measure and record data accurately, completely, and consistently. In other words, are all the bugs worked out so that the system does not lose, change, or misrepresent the data you wish to obtain? I recall, from many years ago, a software program for recording animal weights. If a particular animal had died on study and was not weighed at a weigh session, a "0" was entered for the weight. It was discovered that the software would automatically reject the 0 and record in its place the next animal's weight. This was totally unacceptable. The system was inadequately designed to properly handle commonly occurring data collection exceptions.

The second issue is the verification of the system's operation. Have you tested and proven that the data produced and recorded by the system are accurate, complete, and consistent, meeting all the data quality standards discussed under hand-written data?

Validation and verification are processes that involve hardware and software development, and acceptance testing, laboratory installation procedures and testing, computer security, and special recordkeeping procedures, to name a few. There are numerous publications on this topic. If you are working in a research area subject to FDA or EPA, I suggest starting with the following: the FDA Computerized Data Systems for Nonclinical Safety Assessment — Current Concepts and Quality Assurance, known as the Red Apple Book; the FDA Technical Reference on Software Development Activities; and the EPA's Good Automated Laboratory Practices (GALPs).

The following sections will discuss the defining of raw data for automated data collection systems, what should be recorded in the raw data, electronic signatures, and report formats and spreadsheets.

Computer-generated raw data

It was my privilege to work with the author and a team of experts during the later stages of finalization of the GALPs. One of the most difficult tasks was deciding how to define raw data for laboratory information management systems (LIMS). Hours and days were spent on this issue alone. Here is the definition we ultimately used:

> LIMS Raw Data are original observations recorded by the LIMS
> that are needed to verify, calculate, or derive data that are or
> may be reported. LIMS raw data storage media are the media
> to which LIMS Raw Data are first recorded.

From these discussions, I have developed a broader-based alternative definition of computer-generated raw data. For automated data collection systems, "raw data" mean the first record on the system of original observations that are human readable and that are needed to verify, calculate, or derive data that are or may be reported.

The GALP definition was designed to fit the scope of the GALPs and applicability to EPA's LIMS.

The real issue is how to apply the definition. Hand-recorded raw data is easy to define. What you see is what you write. Automated systems are much more complex. Analytical instruments may perform several functions — a transmitted light beam is measured, is converted into an electronic signal, this signal is transmitted to a computer, the software on the computer converts the signal to a machine-readable representation, this representation is translated into a value, this value is recorded into a report format that performs calculations and a summary of the input data, and the report is sent to an electronic file or to a printer.

The question is when do we have raw data? It is when an understandable value is first recorded. If the human-readable value is saved to a file prior to formatting, this is raw data. If the first recording of the data is in the report format, this is raw data. Some labs have declared the signal from the instrument to the computer to be raw data, but it is then very difficult to use the signal as a means for verification of the report of the data. This example represents only one situation of the possible variations in instrumentation. Each automated data collection system must be assessed to determine when the output is "raw data".

Why is the definition of raw data for computer applications so important? One obvious reason is to meet regulatory requirements. Behind these requirements are the same data quality characteristics that apply to hand-recorded data: accuracy, completeness, consistency, and reconstructability. As mentioned earlier, transcription of data can cause errors. Each time data are translated or reformatted by a software application, there is a potential for the data to be corrupted or lost. When the data are recorded and human readable *before* these operations, these "raw data" can then be used to verify any subsequent iterations.

Here is the type of information that should be included in the automated raw data record:

- The instrument used to collect the data
- The person operating the instrument
- The date (and time) of the operation
- All conditions or settings for the instrument
- The person entering the data (if different from the operator)
- The date and time entered or reported
- The study title or code
- Cross-reference to a notebook or worksheet
- The measurements with associated sample identification
- All system-calculated results

If the system does not allow the input of any of the above information, it may be recorded by hand on the printout or on cross-referenced notebook pages or worksheets.

Automated raw data may be stored in soft copy (e.g., magnetic media) or in hard copy (e.g., paper printout, microfiche, microfilm). However soft copy storage of raw data presents a unique set of problems that are often avoided by printing it in hard copy. Many labs choose to print out raw data, because it assures the data are available and unchanged. More about storage on magnetic media is discussed in chapter three.

Many software applications for instruments record the data in a worksheet format. The same rules as those for hand-generated worksheets should apply for automated formats. However, some raw data may not yet be formatted when they are first recorded. In this case, a key to format of the raw data must accompany the data.

Why do we not designate the final formatted report as raw data in all cases? Remember, in the definition of raw data, the phrase, "first recorded occurrence of the original observation." This is important because the data should have undergone as little manipulation and transfer as possible over different software applications. This prevents corruption and loss and allows the raw data to be used to verify additional operations performed on it. Also, why not designate the signal read by the instrument or transmitted by the instrument as the raw data? Simply because it cannot be understood by humans and therefore is not useful to verify the results and conclusions. Testing should be performed on this signal, however, to validate the operation of the instrument and its communication functions.

Electronic signatures

Electronic signatures are the recorded identity of the individual entering data and are input through log-on procedures — presumed to be secure. One of the issues regarding electronic signatures is the validity of a computer-entered signature because it is not traceable by handwriting analysis to the signer, and presumably anyone could type in a name. One of the changes against Craven Labs (see chapter one Table 1.1) was that the lab changed the clock on the computer to make it appear that samples were analyzed on an earlier date. Currently the FDA is accepting electronically recorded names or initials as signatures although the policy has not been made official at this writing.

Until a policy statement is made, two criteria may be used to justify the use of electronic signatures. All individuals who operate the instruments or associated software must be aware of the meaning and importance of the entry of their name (or unique personal code) and the computerized date stamp. That is what constitutes a legal signature. Second, the electronic signature is best justified when access to the system is strictly controlled. Controlled access usually involves some sort of password or user identification system that must be activated before an authorized person may perform an operation. Some automated systems have levels of access that may control different operations by allowing only certain individuals to perform certain tasks. Access levels may include read only, data entry, data change authorization, and system level entry or change. When these controls are in place, the system may automatically record the persons name into the file based on the password entered. Some systems use voice recognition or fingerprint recognition. This discussion only begins to touch on the complexities of computer security-related issues

Spreadsheets

Spreadsheet use to the modern lab is what invention of the printing press was to publication. Although spreadsheets make recording, processing, and reporting data easy and quick, some special considerations are important to the use of these powerful programs. Whether data are keyed into spreadsheets or electronically transferred to them from existing data files, the entry of the data must be checked to assure the data record is complete and correct. Commonly, mistakes occur in calculations and formulas, in designating data fields, and in performing inappropriate operations on the data. Because of the versatility of spreadsheets, take special care in validating the spreadsheet. When you perform calculations, check the spreadsheet formulas and be sure that the arithmetic formula is defined on the spreadsheet. The way the program rounds numbers and reports significant digits is important to the calculation of results and the reporting of the data. When you try to recalculate or evaluate the processes performed by the spreadsheet program, be sure to define all functions.

Reporting the data

This final section suggests ways to generate data tables and figures for the final report or manuscript. Here are some guidelines:

- The title of the table or figure should be descriptive of the data.
- Column and row headings should be understandable, avoiding undefined abbreviations.
- Units of measure should be included in the column headings or axes of charts.
- For individual data, all missing values must be footnoted and explained.
- All calculations used to derive the data should be defined and, when the calculation is complex or nonstandard, given in a footnote.
- Statistical summaries or analyses should be clearly defined including the type of process performed. Statistically significant values may be identified with a unique symbol that is footnoted.
- All but the most common abbreviations should be defined
- Continuing pages should contain at least a descriptive portion of the title and indicate "continued."
- The data should be easy to read and be uncluttered.
- Charts should contain a legend of any symbols or colors used, and the labels of the axes should be descriptive and easily understood.
- The text of the report should include references to the tables or figures when the data is presented.
- The text of the report should exactly match the data in the tables or figures. Any generalization, summarization, or significant rounding should be designated as such in the text.

Conclusion

The first portion of this chapter was conceptual. The second contained practical implementation methods for handling data. The only way to learn to apply all that was discussed here is through practice. Understanding the principles of recording data comes with experience. To help you begin, here are two application projects for you to try.

Practical applications

Application

Create a project that requires you to take the body weights and measure the food consumption of 50 laboratory rats for 10 days. Develop a simple protocol. Design a form or worksheet to record all the required information. Don't forget to calibrate your balance before and after each weigh session. After the experiment is recorded, design a table showing the data to include in your final report. Note: the animals must be uniquely identified. On Day 3, body weight of one of the animals is incorrectly recorded but not noticed until later. Fix it according to data change guidelines and remember to appropriately report this information.

Alternate application for the practicing researcher

Select a set of existing records. Begin the project by comparing current SOPs to the principles discussed in this chapter and the practices used to record the data. If there are no SOPs in place, create one. Now review that data record. How might it be improved? What errors have you identified? Check some calculations. Record your impressions as you perform this task and discuss them with your colleagues.

References

DeWoskin, R. S. (1995), *Quality Assurance SOPs for GLP Compliance*, Interpharm Press, Buffalo Grove, IL.

Taulbee, S. M. and DeWoskin, R. S. (1993), *Taulbee's Pocket Companion: U.S. FDA and EPA GLPs in Parallel*, Interpharm Press, Buffalo Grove, IL.

U.S. EPA, FIFRA Good Laboratory Practice Standards, Final Rule, *Fed. Reg.* 54:34052–34074, August 17, 1989.

U.S. EPA, TSCA Good Laboratory Practice Standards, Final Rule, *Fed. Reg.* 52:48933–48946, August 17, 1989.

U.S. FDA, Good Laboratory Practice Regulations, Final Rule, *Fed. Reg.* 52:33768–33782, September 4, 1987.

U.S. EPA (1986), Pr Notice 86-5, U.S. EPA, Washington, D.C.

U.S. EPA (1995), Good Automated Laboratory Practices, Office of Information Resources, Research Triangle Park, NC, August 21, 1995.

U.S. FDA, Good Clinical Practices, *CFR*, Title 21, Part 50, 56, 312, 314, April 1, 1993.

U.S. EPA, Enforcement Response Policy for the Federal Insecticide, Fungicide, and Rodenticide Good Laboratory Practices (GLP) Pesticide Enforcement Branch, Office of Compliance Monitoring, Office of Pesticides and Toxic Substances, U.S. EPA.

U.S. EPA, Clean Air Act, Fed. Reg. 59, No. 122, June 27, 1994.

U.S. FDA, Current Good Manufacturing Practices for Finished Pharmaceuticals, *CFR*, Title 21, Part 211.

U.S. FDA, Current Good Manufacturing Practices for Medical Devices, General, *CFR*, Title 21, Part 820.

U.S. FDA, Guide for Detecting Fraud in Bioresearch Monitoring Inspections, Office of Regulatory Affairs, U.S. FDA, April 1993.

U.S. FDA, Electronic Signatures; Electronic Records; Proposed Rule, *Fed. Reg.* 59:13200, August 31, 1994.

Webster's Ninth New Collegiate Dictionary (1983), Merriam-Webster, Springfield, MA.

chapter three

Data retention

Stephanie M. Taulbee

Many laboratory folks think of data retention the same way they view cleaning up after a large holiday meal: tackling the job of organizing and storing study records at the conclusion of the study seems very anticlimactic. Storage and retention of data is, in many cases, mandated by regulation in biomedical research. Further, there may be organizational policies in universities and companies that address retention of records. Contractual agreements also will often detail data retention requirements. There are many reasons for these requirements. Foremost to many researchers is the potential need to refer to the data later, perhaps for additional work being performed or because of inquiries based on publications. To the regulators, however, the records from a study represent legal documentation of the conduct of the study; they are the evidence that supports research results and claims. In patent situations, the records support claims of ownership and discovery.

This chapter will discuss data retention requirements: the facilities required for storage, data and specimen storage methods, and retrieval procedures. The discussion will not be limited to data storage but will cover also the storage of specimens and samples. I refer to the facilities as an archive.

Data retention requirements

What are the retention requirements for the data you are generating? The answer depends on the type of data, the purpose of the data, and the applicable organizational, contractual, and regulatory rules. Table 3.1 suggests some of the policies and regulations that apply to biomedical research. Table 3.2 lists some of the more common requirements. It will be necessary to investigate thoroughly the internal policies of your particular organization, the regulations and contractual agreements for particular research projects, and the available facilities for archiving of data, specimens, and samples.

As you can see, numerous regulatory requirements may apply to data retention. These regulations often contain specifications for the storage facilities or archives, what data are to be retained, and for how long. Fortunately, the FDA and EPA have made their requirements similar except for the length of retention. In addition, FDA and EPA regulations have been used to adopt international standards. Investigate the regulations applicable to your research and follow the most stringent requirements. Table 3.2 defines some of the requirements for facilities, data, and retention time for the major regulations.

Table 3.1 Policies and Regulations for Retention of Data

Research focus	Applicable policy or regulation
University research — thesis or dissertation	University policy and/or grant requirements
University research — contract or grant funded	University policy, grant requirements, NIH, EPA GLP, FDA GLP, FDA GCP, or other[a]
Private industry R & D	Internal policy, patent laws
Pharmaceutical marketing permit	FDA GLPs, cGMPs, GCPs[a]
Pharmaceutical manufacturing	FDA cGMPs[a]
Chemical industry marketing permit for chemicals produced in certain quantities and of certain hazard classes	EPA TSCA GLPs, CAA testing requirements, CWA testing requirements[a]
Pesticide marketing permit	EPA FIFRA GLPs, FDA GLPs[a] for food additives

[a] National Institutes of Health (NIH), Environmental Protection Agency (EPA) Good Laboratory Practice Standards (GLP), Food and Drug Administration (FDA) Good Laboratory Practice Regulations (GLP), FDA Good Clinical Practices (GCP), current Good Manufacturing Practices (cGMP), Toxic Substances Control Act (TSCA), Federal Insecticide, Fungicide, and Rodenticide Act (FIFRA), Clean Air Act (CAA), Clean Water Act (CWA).

Table 3.2 Facility and Retention Requirements of the Major Regulations

Regulation	Facility requirements	Data	Retention requirements
USDA, NIH, and PHS[a] (for animal research facilities)	None specified	IACUC records and reports; training files	Duration of activity plus 3 years
CLIA[a]	None	Records and patient records	2 years
GCPs[a]	None stated, but many follow GLPs	IRB records, case histories, disposition of drug reports	2 years following acceptance of the report by the agency
GLPs[a]	Restricted access	All raw data documentation records, protocols, specimens, and final reports	(See Table 3.4) FDA — 2 years EPA TSCA — 5 to 10 years EPA FIFRA — 2 years, 5 years, or the lifetime of the registration of the chemical
FDA cGMP[a]	Per GLP	Per GLP	Per GLP

[a] U.S. Department of Agriculture (USDA), National Institutes of Health (NIH), Public Health Service (PHS), Clinical Laboratory Improvement Act (CLIA), Good Clinical Practices (GCP), Institutional Animal Care and Use Committee (IACUC), Institutional Review Board (IRB).

Academic research

Academic research may not fall under any of the above regulations or guidelines. However, institutional policies will often require retention of data for a period of time following completion of defense of a thesis or dissertation, publication of the research results, or at the end of a research project. For example, some universities require that data be retained for 2 years following defense of a thesis or dissertation and that the thesis or dissertation be retained in the university library indefinitely. Commonly, there are no specific requirements for what data must be retained and no facilities are provided for retention of the data. Departmental policies may be established that respond to these uncertainties and to address budgetary and space constraints. A minimal policy may state that all research data be gathered at the end of a project, inventoried, and filed in locked file cabinets, with the inventory and location of the data filed in the department office. Departments may also designate a locked storage room for data storage. Finally, if your research requires a formal archives, then secure limited-access storage facilities will have to be established. The remaining sections of this chapter will address regulatory requirements and how these requirements are put into practice. Review of these sections may provide useful suggestions that may be applied or modified slightly when developing a data retention policy in an academic setting.

Regulatory requirements

The requirements for GLP archives are described because they are the most specific of the federal research regulations. However, this chapter does not address regulations concerning classified documents. Each section provides a paraphrase of the GLP requirement and then discusses suggested ways to implement the requirement.

Archive facilities

Archives are to be secure with established SOPs that limit access to authorized personnel and that assign a responsible person to manage the archives. The facilities must provide proper storage conditions for the maintenance of records and specimens. The purpose of the archives is to provide orderly storage and retrieval of the study records and specimens. This section will describe the regulatory requirements and provide discussion and practical suggestions for compliance with each requirement.

Requirement — There shall be archives for orderly storage and expedient retrieval of all raw data, documentation, protocols, specimens, and interim and final reports.

Discussion — The two key points are the orderly storage and the retrieval procedures. Procedures must be in place for organizing, boxing, and inventorying records and specimens. Then, there should be procedures for organization of the boxes of data and specimens and transfer to the archives. An archive form may be used that provides the archivist with information about the study and the contents of the boxes (see Figure 3.1). This form is submitted when data are transferred to the archives and documents the transfer. The archivist will then inventory the boxes and receive them into the archives.

For archives that contain several studies, you will need a system of recording the location and contents of the boxes so that they may then be retrieved easily. Retrieval may be required by the investigator or by auditors from the sponsor or regulatory agencies. This information may be recorded on the archive form, and the form may be filed by study designation or investigator. Many facilities have established databases that record the study, the investigator, the number of boxes and their contents, and the location of the boxes in the archives. The database may be a reiteration of the information specified on the form, which allows the archivist to search different fields in the database when retrieval is required.

Retrieval procedures must be established. Only certain individuals should be allowed to request that data or specimens be removed from the archives: the inves-

ARCHIVE INVENTORY

Please complete all blanks

Total Number of Boxes: ____________________

Laboratory: ____________________

Sponsor and Contract Number: ____________________

Study Code Number: ______________ Protocol Number: ______________

Type of Study: ____________________

Study Chemical/Test Article: ____________________

Animal/Test System: ____________________

Applicable GLPs: FDA ____ EPA-FIFRA ____ EPA-TSCA ____ Non-GLP ____

Type of Submission (GLP Studies Only) ____________________

Proposed Removal Date ____________________

Please sign below

____________________ ___/___/___

P.I. or Authorized Lab. Personnel — Date

For Archivist Use

Entry Date ______________ Alphanumeric Location ______________

Total Number of Boxes: ______________ ______________ ___/___/___

Archivist — Date

List a brief inventory of the contents of each box submitted for storage

Box Number	Description of Contents in Each Box

(continue on additional pages, if needed)

Page ____ of ____

Figure 3.1 Archive inventory form.

tigator may be granted access to the records when research is ongoing; auditors from the sponsor or regulatory agencies may have access to the records. Other individuals must have permission of the investigator or upper management to gain access to data. When access is requested, some organizations require that the records not be removed from the archives but be reviewed within the confines of the archive facility. Others will sign out data to authorized personnel and maintain records of the removal. These records should indicate the person to whom the records are given, their location, and expected time of return. When the records are returned, the documentation includes the date of the return and the new location of the boxes applicable. It is essential to maintain the chain of custody of the boxes or contents.

Requirement — Conditions of storage shall minimize deterioration of the documents or specimens in accordance with the requirements for the time period of their retention and the nature of the documents or specimens.

Discussion — The archives must be designed to meet minimum storage conditions for whatever type of data media or specimens that are to be stored. In most cases, paper data, formalin-fixed wet tissue, slides, and paraffin or plastic tissue blocks may be stored at normal room temperature, 72 ± 3°F, with normal room humidity of 30 to 70% RH. However, some types of data storage media and methods of specimen fixation will require special conditions. Table 3.3 describes different storage requirements for data and specimens along with the expected life under the conditions.

Table 3.3   Retention Conditions Per Record Storage Medium

Records	Storage conditions	Shelf life under specified conditions
Paper records	Normal room temperature and humidity	
Light-sensitive records	Protection from all light sources	
Heat-sensitive records	Refrigeration or coldroom facilities	
Magnetic media — diskette	Normal room temperature and humidity, away from magnetic fields	Manufacturer specifications / may be 10 years; must maintain computer and software to read
Magnetic media — compact disk	Normal room temperature and humidity, away from magnetic fields	Estimated at 100 years; must maintain equipment and software to read
Magnetic media — tapes	Normal room temperature and humidity, away from magnetic fields	Exercised on an annual cycle and maintain software and hardware to read
Photographic film and prints	Cool dark	10 years
Tissues by common fixative type		
Formalin[a]	Normal room temperature and humidity	10 years or more
Alcohol/glycerin[a]	Tightly sealed; room temperature and humidity	10 years or more
Glutaraldehyde/osmium tetroxide[a]	Do not store tissues in these fixatives	
Picric acid (Bouin's)[a]	Specimens should be transferred to another fixative (70% alcohol)	Use retention time for chosen fixative
Frozen	–20 or –40°C	Depends on the analysis endpoint

[a] Sheehan and Hrapchak (1980).

Requirement — A testing facility may contract with commercial archives to provide a repository for all material to be retained. Raw data and specimens may be retained elsewhere, provided that the archive records make specific reference to those other locations.

Discussion — This requirement allows a testing facility to establish a contractual agreement with a commercial archive for the retention of the study materials. Implicit in the requirement is that the commercial archives provide proper facilities and maintain the necessary documentation required by other sections of the regulations. The testing facility must know where all the materials are located, and the archive records must contain documentation of the whereabouts of these materials. To accomplish this, the testing facility should designate a responsible person or archivist to keep track of the materials, even when the facilities are offsite or at a commercial archive. The commercial archive will maintain documentation of any raw data or specimens that may be located elsewhere. The purpose is to cross-reference at all locations the whereabouts of all records and specimens related to the study.

Requirement — Material retained or referred to in the archives shall be indexed to permit expedient retrieval.

Discussion — The location of the raw data, specimens, and final report must be identified in the final report for GLP studies submitted to the agency. Contract laboratories may retain custody of the data or turn custody over to the sponsor. Procedures for this change of custody must be clearly established, with documentation maintained by the testing facility. Regulatory agencies require that the testing facility produce all materials for an audit unless an alternate location for the data is specified in the archive records. Beyond the method of indexing records, using a database helps the archivist maintain the necessary records.

Study management responsibilities

Requirement — The study director shall assure that all raw data, documentation, protocols, specimens, and final reports are transferred to the archives during or at the close of the study.

Discussion — The responsibility for assuring that all required materials are transferred to the archives is clearly placed on the study director or investigator. These materials must be transferred to the archives during or at the close of the study. The close of the study is when the final report is signed. In reality, the timing of signatures is often within hours or minutes of final changes to that report, requiring access to at least some of the raw data up to the last minute. This may not be the best scenario but it is reality. It is therefore recommended that the data and specimens be prepared for archiving, organized, and boxed, during the report-writing stage. Then the final inventory and transfer can take place immediately following the final sign-off of the report. How the regulators interpret the phrase "during or at the close of the study" is not clear. Does that mean the same day, or is there a grace period of a week or a month or longer? Clearly, the same day meets the most stringent interpretation. If facilities find it necessary to allow longer periods of time, then they must be prepared to defend the allowances and should clearly document their internal policy in the form of an SOP. Theoretically, the reason for this GLP requirement is to protect the data from changes or loss after the study is officially complete. In fact, the reason for all the archive requirements is to protect the study materials (data and records) from change or loss. This concept should drive any related policy by the organization.

The Archivist's responsibilities

Requirement — An individual shall be identified as responsible person for the archives.

Discussion — This individual is often referred to as the archivist. While a single individual is required, most companies designate an alternate who is authorized by the archivist to substitute for the archivist in the event of an absence. There is no specific requirement for the archivist to be independent from the research staff, but some organizations give this responsibility to someone in the Quality Assurance Unit (QAU). The QAU is often the logical choice for control of the archives because it works with the records, with the investigator, to insure that records and specimens are archived at the end of the study, and often is given the responsibility of hosting audits by the sponsor and regulatory agencies. In other cases, the archivist or archive staff are organizationally separate from the research or QAU staff, in which case the QAU will work closely with the archivist to assure all applicable requirements are met.

Data and specimen retention

This section discusses what must be retained and describes some of the considerations in retaining data and specimens. The GLPs have specific requirements for data generated during the conduct of a safety assessment or nonclinical study. There are also special requirements for clinical studies. And there are additional requirements for animal research approvals and for chemical handling and disposal records. Finally, there may be requirements for retention of financial information and records within a particular institution.

GLP retention requirements

Requirement — All raw data, documentation, records, protocols, specimens, and final reports generated as a result of a study shall be retained.

Discussion — What does this include? The answer is fairly straightforward. Every piece of information associated with the technical conduct of the study, with the exception of draft reports, is part of the study record and must be retained. The raw data, notes, correspondence, receipts, the protocol and amendments, and final report are included in the study record. Associated equipment use and calibration logs, maintenance records, and personnel training files may not be study-specific records but also must be retained and archived.

What else is there, you may ask? There are no retention requirements for financial records in the GLPs. Financial records include contractual agreements, budget and payment records, correspondence dealing with financial considerations, and personnel employment and salary records. These records must be retained for other purposes, but are not study records and should not be included in the GLP archive files.

The following is a partial list of the records to be retained for a preclinical animal study.

- Protocol and amendments
- Test, control, and reference substance receipt
- Test system receipt
- Quarantine of test system
- Test system storage and use
- Animal facility use maintenance and calibration records
  - Temperature and humidity records
  - Cage and cage rack cleaning and washer control documentation
  - Animal feed and bedding receipt and storage
- Analysis of bulk test, control, and reference substance for identity, strength, purity, stability, and uniformity
- Dose preparation and analysis for strength, stability, and homogeneity
- Dose administration records

- Test data such as body weights, food and water consumption, clinical observations, and evaluations
- Animal husbandry, including cage changes and animal room cleaning documentation
- Unused animal disposition
- Necropsy evaluation and collection of specimens
- Processing and evaluation of specimens
- Data entry and evaluation records
- Statistical analyses
- Final report

Specimen retention

Requirement — FDA GLPs state that specimens (except those obtained from mutagenicity tests and wet specimens of blood, urine, feces, and biological fluids) generated as a result of a nonclinical laboratory study shall be retained.

EPA GLPs indicate the following: specimens obtained from mutagenicity tests; specimens of soil, water, and plants; and wet specimens of blood, urine, feces, and biological fluids do not need to be retained after quality assurance verification.

Wet specimens, samples of test and/or control articles, and specially prepared materials that are relatively fragile and differ markedly in stability and quality during storage shall be retained only as long as the preparation affords evaluation.

Discussion — The purpose of this requirement is to retain all specimens that may be evaluated at some time within the retention time requirements and that collaborate raw data and conclusions of the study. Certain types of wet specimens or perishable specimens have been excluded. The criteria for the exclusion is whether the specimen may be reevaluated after storage. Here, a caution should be issued. While whole blood may not be reevaluated after only a day or so for a complete blood count, serum for serum analysis for chemical levels may be retained frozen indefinitely and may be reanalyzed. The determination of whether a sample specimen must be retained depends on the type of analysis, the sample medium, and the storage conditions. Since the GLPs require that storage conditions must be those that minimize deterioration, the conditions that allow for reanalysis must be used, if possible. It may *not* be assumed that all blood, urine, and feces may be automatically discarded.

The EPA requires that the QAU verify the disposal of specimens, which is interpreted to mean that the QAU must have direct knowledge of the existence of the specimens and be provided with a justification for disposing of them. The justification should include a reasonable scientific explanation of why the specimens are no longer useful for reevaluation. This may require the scientist to have documentation of historical information or actual reanalysis data documenting that the specimens cannot be adequately analyzed. However, the QAU is generally not qualified to assess these types of scientific decisions and must rely on the scientist's best judgment. For EPA there must be documentation of the QAU verification of all samples that are discarded.

QAU records

Requirement — The GLPs require that the certain records be maintained by the QAU. While there is no formal requirement for archiving these records, most laboratories will place the QAU records in the archives for the period of time specified for the study records. The QAU records that must be maintained are the master schedule of all studies, copies of protocols and amendments, copies of inspection reports, copies of audit reports, reports to management, and the QAU SOPs. These records should not be retained in the study files. Only portions of the QAU records are required to be reviewed by designated representatives of the regulatory agencies. The GLPs state:

> The responsibilities and procedures applicable to the quality assurance unit, the records maintained by the quality assurance unit, and the method of indexing such records shall be in writing and shall be maintained. These items including inspection dates, the study inspected, the phase or segment of the study inspected and the name of the individual performing the inspection shall be made available for inspection to authorized employees or duly designated representatives of the EPA or FDA.

Discussion — In the second sentence, "these items" refers to written documentation of the responsibilities and procedures, not the actual records.

Requirement — The GLPs further state, "An authorized employee or a duly designated representative of the EPA or FDA shall have access to the written procedures and may request testing facility management to certify that inspections are being implemented, performed, documented and followed up in accordance with this paragraph."

Discussion — The regulatory agency officials have stated that the purpose of this policy is to provide open and free internal communication between the QAU and facility management and the study director. You can imagine what the records would contain if federal investigators had access to all the QAU records. The simple truth is that any findings made by the QAU and the corrective actions taken that apply to the conduct of the study are documented in the study records by the existence of a properly documented study record and corrections made to that record. The study records speak for themselves and the proper functioning of the QAU.

The actual QAU records are maintained as required and are often archived, but are available only by subpoena to demonstrate, in the event of a criminal or civil action, that the QAU functioned as required.

The regulatory agencies have recently increased their emphasis on the sponsor's responsibility for the GLP compliance of contract facilities. Because of this, some sponsors' representatives may ask to see QAU inspection reports and reports to management for their studies. The contract organization must decide whether to allow this or whether to provide assurances to the sponsor that may satisfy their need to know. In deciding on this issue, two things must be considered. Consider first the internal communication channels and whether allowing clients and sponsors the privilege of inspecting these communications would alter their effectiveness. Also, consider the protection of confidentiality of information regarding other clients or sponsors which may be dealt with by restricting access only to records for the particular sponsor or study and by purging all joint records or reports or any reference to other sponsors or studies. My professional opinion is that sponsors may receive adequate evidence of GLP compliance during an audit or site visit without direct access to QAU records.

Retention of records

Requirements — Table 3.4 describes the various time requirements cited in the FDA and EPA GLPs. Some labs may perform work under a variety of regulations with different retention times. Some organizations may require that data and specimens be retained for longer periods of time specified in the federal requirements. The procedures used to determine the retention times and disposal procedures must be clearly documented in facility SOPs. A court ruling held A. H. Robbins, makers of the Dalkon Shield, liable for retention of study data in support of safety claims, even though disposal of the data was carried out after the retention requirements of the FDA. Behind the court's ruling was the fact that Robbins did not have a standard procedure for discarding data. The ruling was designed to prevent industry from discarding incriminating data when there is impending legal action against them.

Discussion — The minimum retention requirements for all but fragile specimens may be 2 years, or can be as long as the lifetime of the registration of the chemical. Therefore, it is important to determine for each study what regulations apply and their associated retention requirements. This is often difficult for contract facilities. The time when disposal is permitted depends on the date of submission of the data to the agency in most cases. Information about the disposition of the submission may not be available from the sponsoring company. Therefore, the contract lab should adopt a policy that requests permission to dispose of data and specimens when they think the requirements are met. This will often elicit an appropriate response by the sponsor to dispose of them or retain them for an additional period. Some sponsors will request that data and/or specimens be transferred to them at that time. Appropriate approvals and documentation must be maintained by the testing facility archives when transfer is made. Procedures should be described in the SOPs for this transfer.

Retention of test, control, and reference substance reserve samples

Requirement — For studies of more than 4 weeks experimental duration, reserve samples from each batch of test, control, and reference substances shall be retained for the period of time provided by EPA TSCA GLP Part 160.195 (also FIFRA GLP Part 792.195 or FDA GLP Part 58.195).

Discussion — For studies with an experimental duration of more that 4 weeks, reserve samples of the bulk and all batches of dose formulation must be retained [found in 58.1067, 160.105(d), and 792.105(d)]. "Four weeks" refers to the dosing period, not the entire period from the signing of the protocol to the final report. This requirement refers only to the retention period specified in the applicable GLPs, which is where the retention times are discussed. While it does not specify that these samples must be kept in the archives, some formal mechanism should be established for maintaining these samples. They should be maintained using the archive criteria: secure facilities (locked with limited access), store to prevent deterioration, maintain documentation of their location, and record any transfers of the samples to another location. Also, if the samples are not stable beyond a known and documented time period, they do not have to be retained beyond that period. It is required that all samples have an expiration date on them, if appropriate. This will facilitate disposal at appropriate times. Sponsors may require contract labs to request permission to dispose of these samples.

Table 3.4   FDA and EPA GLP Regulations

21 CFR Part 58 FDA Good Laboratory Practice for Nonclinical Laboratory Studies (Revised) Effective Date: October 5, 1987	40 CFR Part 160 EPA Federal Insecticide, Fungicide and Rodenticide Act (FIFRA); Good Laboratory Practice Standards; Final Rule: August 17, 1989	40 CFR Part 792 EPA Toxic Substances Control Act (TSCA); Good Laboratory Practice Standards; Final Rule; August 17, 1989
58.190 *Storage and retrieval of records and data.* (a) All raw data, documentation, protocols, final reports, and specimens (except those specimens obtained from mutagenicity tests and wet specimens of blood, urine, feces, and biological fluids) generated as a result of a nonclinical laboratory study shall be retained.	160.190 *Storage and retrieval of records and data.* (a) All raw data, documentation, records, protocols, specimens, and final reports generated as a result of a study shall be retained. Specimens obtained from mutagenicity tests, specimens of soil, water, and plants, and wet specimens of blood, urine, feces, and biological fluids, do not need to be retained after quality assurance verification. Correspondence and other documents relating to interpretation and evaluation of data, other than those documents contained in the final report, also shall be retained.	792.190 *Storage and retrieval of records and data.* (a) All raw data, documentation, records, protocols, specimens, and final reports generated as a result of a study shall be retained. Specimens obtained from mutagenicity tests, specimens of soil, water, and plants, and wet specimens of blood, urine, feces, and biological fluids, do not need to be retained after quality assurance verification. Correspondence and other documents relating to interpretation and evaluation of data, other than those documents contained in the final report, also shall be retained.
(b) There shall be archives for orderly storage and expedient retrieval of all raw data, documentation, protocols, specimens, and interim and final reports. Conditions of storage shall minimize deterioration of the documents or specimens in accordance with the requirements for the time period of their retention and the nature of the documents or specimens. A testing facility may contract with commercial archives to provide a repository for all material to be retained. Raw data and specimens may be retained elsewhere, provided that the archives have specific reference to those other locations. (c) An individual shall be identified as responsible for the archives. (d) Only authorized personnel shall enter the archives. (e) Material retained or referred to in the archives shall be indexed to permit expedient retrieval.	(b) There shall be archives for orderly storage and expedient retrieval of all raw data, documentation, protocols, specimens, and interim and final reports. Conditions of storage shall minimize deterioration of the documents or specimens in accordance with the requirements for the time period of their retention and the nature of the documents or specimens. A testing facility may contract with commercial archives to provide a repository for all material to be retained. Raw data and specimens may be retained elsewhere provided that the archives have specific reference to those other locations. (c) An individual shall be identified as responsible for the archives. (d) Only authorized personnel shall enter the archives. (e) Material retained or referred to in the archives shall be indexed to permit expedient retrieval.	(b) There shall be archives for orderly storage and expedient retrieval of all raw data, documentation, protocols, specimens, and interim and final reports. Conditions of storage shall minimize deterioration of the documents or specimens in accordance with the requirements for the time period of their retention and the nature of the documents or specimens. A testing facility may contract with commercial archives to provide a repository for all material to be retained. Raw data and specimens may be retained elsewhere provided that the archives have specific reference to those other locations. (c) An individual shall be identified as responsible for the archives. (d) Only authorized personnel shall enter the archives. (e) Material retained or referred to in the archives shall be indexed to permit expedient retrieval.
58.195 *Retention of records.* (a) Record retention requirements set forth in this section do not supersede the record retention requirements of any other regulations in this chapter. (b) Except as provided in paragraph (c) of this section, documentation records, raw data and specimens pertaining to a nonclinical laboratory study and required to be made by this part shall be retained in the archive(s) for whichever of the following periods is shortest:	160.195 *Retention of records.* (a) Record retention requirements set forth in this section do not supersede the record retention requirements of any other regulations in this subchapter. (b) Except as provided in paragraph (c) of this section, documentation records, raw data and specimens pertaining to a study and required to be made by this part shall be retained in the archive(s) for whichever of the following periods is longest:	792.195 *Retention of records.* (a) Record retention requirements set forth in this section do not supersede the record retention requirements of any other regulations in this subchapter. (b)(1) Except as provided in paragraph (c) of this section, documentation records, raw data and specimens pertaining to a study and required to be retained by this part shall be retained in the archive(s) for a period of at least 10 years following the effective date of the applicable final test rule.

Table 3.4 (continued)   FDA and EPA GLP Regulations

21 CFR Part 58 FDA Good Laboratory Practice for Nonclinical Laboratory Studies (Revised) Effective Date: October 5, 1987	40 CFR Part 160 EPA Federal Insecticide, Fungicide and Rodenticide Act (FIFRA); Good Laboratory Practice Standards; Final Rule; August 17, 1989	40 CFR Part 792 EPA Toxic Substances Control Act (TSCA); Good Laboratory Practice Standards; Final Rule; August 17, 1989
(1) A period of at least 2 years following the date on which an application for a research or marketing permit, in support of which the results of the nonclinical laboratory study were submitted, is approved by the Food and Drug Administration. This requirement does not apply to studies supporting notices of claimed investigational exemption for new drugs (INDs) or applications for investigational device exemptions (IDEs), records of which shall be governed by the provisions of paragraph (b) (2) of this section.	(1) In the case of any study used to support an application for a research or marketing permit approved by EPA, the period during which the sponsor holds any research or marketing permit to which the study is pertinent.	
(2) A period of at least 5 years following the date on which the results of the nonclinical laboratory study are submitted to the Food and Drug Administration in support of an application for a research or marketing permit.	(2) A period of at least 5 years following the date on which the results of the study are submitted to the EPA in support of an application for a research or marketing permit.	(2) In the case of negotiated testing agreements, each agreement will contain a provision that, except as provided in paragraph (c) of this section documentation records, raw data, and specimens pertaining to a study and required to be retained by this part shall be retained in the archive(s) for a period of at least ten years following the publication date of the acceptance of a negotiated test agreement.
(3) In other situations (e.g., where the nonclinical laboratory study does not result in the submission of the study in support of an application for a research or marketing permit), a period of at least 2 years following the date on which the study is completed, terminated, or discontinued.	(3) In other situations (e.g., where the study does not result in the submission of the study in support of an application for a research or marketing permit), a period of at least 2 years following the date on which the study is completed, terminated, or discontinued.	(3) In the case of testing submitted under Section 5, except for those items listed in paragraph (c) of this section, documentation records, raw data, and specimens pertaining to a study and required to be retained by this part shall be retained in the archive(s) for a period of at least 5 years following the date on which the results of the study are submitted to the agency.
(c) Wet specimens (except those specimens obtained from mutagenicity tests and wet specimens of blood, urine, feces, and biological fluids), samples of test or control articles, and specially prepared material which are relatively fragile and differ markedly in stability and quality during storage, shall be retained only as long as the quality of the preparation affords evaluation. In no case shall retention be required for longer periods than those set forth in paragraphs (a) and (b) of this section.	(c) Wet specimens, samples of test, control, or reference substances, and specially prepared material which are relatively fragile and differ markedly in stability and quality during storage, shall be retained only as long as the quality of the preparation affords evaluation. Specimens obtained from mutagenicity tests, specimens of soil, water, and plants, and wet specimens of blood, urine, feces, and biological fluids, do not need to be retained after quality assurance verification. In no case shall retention be required for longer periods than those set forth in paragraphs (b) of this section.	(c) Wet specimens, samples of test, control, or reference substances, and specially prepared material which are relatively fragile and differ markedly in stability and quality during storage, shall be retained only as long as the quality of the preparation affords evaluation. Specimens obtained from mutagenicity tests, specimens of soil, water, and plants, and wet specimens of blood, urine, feces, and biological fluids, do not need to be retained after quality assurance verification. In no case shall retention be required for longer periods than those set forth in paragraph (b) of this subsection.

(d) The master schedule sheet, copies of protocols, and records of quality assurance inspections, as required by 58.35(c) shall be maintained by the quality assurance unit as an easily accessible system of records for the period of time specified in paragraphs (a) and (b) of this section.

(e) Summaries of training and experience and job descriptions required to be maintained by 58.29(b) may be retained along with all other testing facility employment records for the length of time specified in paragraphs (a) and (b) of this section.

(f) Records and reports of the maintenance and calibration and inspection of equipment, as required by 58.63(b) and (c), shall be retained for the length of time specified in paragraph (b) of this section.

(g) Records required by this part may be retained either as original records or as true copies such as photocopies, microfilm, microfiche, or other accurate reproductions of the original records.

(h) If a facility conducting nonclinical testing goes out of business, all raw data, documentation, and other material specified in this section shall be transferred to the archives of the sponsor of the study. The Food and Drug Administration shall be notified in writing of such a transfer.

(d) The master schedule sheet, copies of protocols, and records of quality assurance inspections, as required by 160.35(c) shall be maintained by the quality assurance unit as an easily accessible system of records for the period of time specified in paragraph (b) of this section.

(e) Summaries of training and experience and job descriptions required to be maintained by 160.29(b) may be retained along with all other testing facility employment records for the length of time specified in paragraph (b) of this section.

(f) Records and reports of the maintenance and calibration and inspection of equipment, as required by 160.63(b) and (c), shall be retained for the length of time specified in paragraph (b) of this section.

(g) If a facility conducting testing or an archive contracting facility goes out of business, all raw data, documentation, and other material specified in this section shall be transferred to the archives of the sponsor of the study. The EPA shall be notified in writing of such a transfer.

(h) Specimens, samples, or other non-documentary materials need not be retained after EPA has notified in writing the sponsor or testing facility holding the materials that retention is no longer required by EPA. Such notification normally will be furnished upon request after EPA or FDA has completed an audit of the particular study to which the materials relate and EPA has concluded that the study was conducted in accordance with this part.

(i) Records required by this part may be retained either as original records or as true copies such as photocopies, microfilm, microfiche, or other accurate reproductions of the original records.

(d) The master schedule sheet, copies of protocols, and records of quality assurance inspections, as required by 792.35 (c) shall be maintained by the quality assurance unit as an easily accessible system of records for the period of time specified in paragraph (b) of this section.

(e) Summaries of training and experience and job descriptions required to be maintained by 792.29 (b) may be retained along with all other testing facility employment records for the length of time specified in paragraph (b) of this section.

(f) Records and reports of the maintenance and calibration and inspection of equipment, as required by 792.63 (b) and (c), shall be retained for the length of time specified in paragraph (b) of this section.

(g) If a facility conducting testing or an archive contracting facility goes out of business, all raw data, documentation, and other material specified in this section shall be transferred to the archives of the sponsor of the study. The EPA shall be notified in writing of such a transfer.

(h) Specimens, samples, or other non-documentary materials need not be retained after EPA has notified in writing the sponsor or testing facility holding the materials that retention is no longer required by EPA. Such notification normally will be furnished upon request after EPA or FDA has completed an audit of the particular study to which the materials relate and EPA has concluded that the study was conducted in accordance with this part.

(i) Records required by this part may be retained either as original records or as true copies such as photocopies, microfilm, microfiche, or other accurate reproductions of the original records.

Clinical studies retention requirements

Requirements — Clinical studies (with human subjects) conducted under the FDA GCPs must retain certain types of documentation for defined periods of time. While the regulations do not include specific requirements for archival storage, facilities have adopted the GLP facility requirements. Table 3.5 lists the types of data that must be retained and the period of time required.

Institutional animal care and use committee records

The Institutional Animal Care And Use Committee (IACUC) is charged with the responsibility of overseeing animal care and use in research projects. Therefore, it must maintain documentation of its activities and submit reports to the governing federal agencies. Because of this, there are data retention requirements from the PHS, the USDA, FDA, and EPA that govern retention of records. The institution must identify the requirements and write institutional policies and SOPs for retention of records according to the applicable laws.

Conclusion

This chapter has been designed to introduce to you the major requirements for records and specimen retention encountered by biomedical researchers; it has not been comprehensive. All regulated data collection is accompanied by implicit and explicit data retention requirements. I have not addressed clinical research performed for purposes other than market permit application, nor have I discussed epidemiologic research. I have, also, not addressed data retention requirements of hospitals or clinical laboratories. In reality, this chapter provided only a taste of the regulations that might apply in different research situations. Again, it is the responsibility of investigators to find out what requirements apply to their projects and to prescribe procedures for compliance with those requirements.

Practical applications

Application

The assignment for this chapter is to gather the information on the internal policies for the work you are currently performing. If you are a graduate student, determine the requirements of the university with regard to retention of data and the requirements of any grant that is funding your work. If you are a researcher, determine internal policies and regulatory requirements. Get copies of any SOPs that apply to your work.

Alternate assignment

Contact a researcher about a study completed in the last few years and determine what was done with the data, their location, the requirements for retention of the data, and whether they were followed.

Table 3.5   Requirements for Human Subject Research (From FDA
Good Clinical Practices)

Data retention requirement as stated in the FDA Good Clinical Practice regulations	Retention time

Institutional Review Board (IRB) Records:
  1. Copies of all research proposals reviewed, scientific evaluations, if any, that accompany the proposals, approved sample consent documents, progress reports submitted by investigators, and reports of injuries to subjects
  2. Minutes of IRB meetings which shall be in sufficient detail to show attendance at the meetings, actions taken by the IRB, the vote on these actions including the number of members voting for, against, and abstaining: the basis for requiring changes in or disapproving research: and a written summary of the discussion of controverted issues and their resolution
  3. Records of continuing review activities
  4. Copies of correspondence between the IRB and the investigators
  5. A list of IRB members identified by name, earned degrees, representative capacity, indication of experience such as board certifications, licenses, etc.; sufficient to describe each members chief anticipated contributions to IRB deliberations; and any employment or other relationship between each member and the institution
  6. Written procedures for the IRB as required by 58.108(a)
  7. Statements of significant new findings provided to subjects as required by 50.25
	3 years following completion of the research

Sponsor record requirements A sponsor shall maintain adequate records showing the receipt, shipment, or other disposition of the investigational drug. These records are required to include, as appropriate, the name of the investigator to whom the drug is shipped and the date, quantity, and batch or code of each such shipment	2 years following approval of the market permit or, if not approved, 2 years after the last shipment of the drug to the investigator

Investigator record requirements
  The investigator is required to maintain adequate records of the disposition of the drug including dates, quantity, and use by subjects. If the investigation is terminated, suspended, discontinued or completed, the investigator shall return the unused supply of the drug to the sponsor, or otherwise provide for disposition of the unused supplies of the drug under 312.59
  Case histories. An investigator is required to prepare and maintain adequate and accurate case histories designed to record all observations and other data pertinent to the investigation on each individual treated with the investigational drug or employed as a control in the investigation
	Not specifically stated

Table 3.5 (continued) Requirements for Human Subject Research (From FDA Good Clinical Practices)

Data retention requirement as stated in the FDA Good Clinical Practice regulations	Retention time
Reports	
a. Progress Reports The investigator shall furnish all reports to the sponsor of the drug who is responsible for collecting and evaluating the results obtained. The sponsor is required under sec. 312.33 to submit annual reports to FDA on the progress of the clinical investigations.	Implied that all reports are retained as other records shown above since FDA investigators have access to these reports in an audit
b. Safety Reports. An investigator shall promptly report to the sponsor any adverse effect that may reasonably be regarded as caused by, or probably caused by, the drug. If the adverse effect is alarming, the investigator shall report the adverse effect immediately.	
c. Final Report. An investigator shall provide the sponsor with an adequate report shortly after completion of the investigator's participation in the investigation.	

References

DeWoskin, R. S, (1995), *Quality Assurance SOPs for GLP Compliance*, Interpharm Press, Buffalo. Grove, IL.

Guide to the Care and Use of Laboratory Animals, National Institutes of Health, Bethesda, MD, 1985–1986.

Institutional Animal Care and Use Committee Guidebook, NIH Pub. No. 92-3415, U.S. Department of Health and Human Services, pp. E33–37.

Sheenhan, D. C. and Hrapchak, B. B. (1980), *Theory and Practice of Histotechnology*, C.V. Mosby, St. Louis.

Taulbee, S. M. and DeWoskin, R. S. (1993), *Taulbee's Pocket Companion: U.S. FDA and EPA GLPs in Parallel*, Interpharm Press, Buffalo Grove, IL.

U.S. EPA, FIFRA Good Laboratory Practice Standards, Final Rule, *Fed. Reg.* 54:34052–34074, August 17, 1989.

U.S. EPA, TSCA Good Laboratory Practice Standards, Final Rule, *Fed. Reg.* 52:48933–48946, August 17, 1989.

U.S. EPA (1986) Pr Notice 86-5, U.S. EPA, Washington, D.C.

U.S. EPA (1995), Good Automated Laboratory Practices, Office of Information Resources, Research Triangle Park, NC, August.

U.S. EPA, Enforcement Response Policy for the Federal Insecticide, Fungicide, and Rodenticide Good Laboratory Practices (GLP), Pesticide Enforcement Branch, Office of Compliance Monitoring, Office of Pesticides and Toxic Substances.

U.S. EPA, Clean Air Act, *Fed. Reg.* 59, No. 122, June 27, 1994.

U.S. FDA, Current Good Manufacturing Practices for Finished Pharmaceuticals, *CFR*, Title 21, Part 211.

U.S. FDA, Good Laboratory Practice Regulations, Final Rule, *Fed. Reg.* 52:33768–33782, September 4, 1987.

U.S. FDA, Good Clinical Practices, *CFR*, Title 21, Part 50, 56, 312, 314, April 1, 1993.

U.S. FDA, Current Good Manufacturing Practices for Medical Devices, General, *CFR*, Title 21, Part 820.

U.S. FDA, Guide for Detecting Fraud in Bioresearch Monitoring Inspections, Office of Regulatory Affairs, U.S. FDA, April 1993.

U.S. FDA, Electronic Signatures; Electronic Records; Proposed Rule, *Fed. Reg.* 59:13200, August 31, 1994.

chapter four

Implementing quality systems in biomedical research

Stephanie M. Taulbee

Quality is by definition a characteristic with respect to excellence or grade of excellence. Chapter one discussed the characteristics that broadly define quality in research and where researchers have erred, and chapter two focuses on incorporating these characteristics into the data. This chapter will describe systems that assure that the quality characteristics are instilled in the data. The objective of this chapter is to introduce you to these concepts and the subtleties inherent in them.

Chapter one demonstrates to us that quality and sometimes honesty are not to be automatically assumed. All the complexities of human nature and human ingenuity contribute to the need for monitoring research practices. Government agencies have recognized this need, particularly in areas that affect the public health. They have established through regulatory mandates systems of quality control (QC) and quality assurance (QA). QC is a system by which controls — internal to the research process — are established, thereby ensuring the quality of the product or results. QA is the process by which assurance is given to someone with a vested interest that established standards of quality have been achieved in some product. In the GLP arena, QA means assuring laboratory management that, during the conduct of a regulated study, the protocol, SOPs, and GLPs are followed and that the reported results accurately reflect the raw data. More broadly, the scientific community, the public (individually and collectively), our government agencies, sponsoring organizations, and laboratory management all have a vested interest in the quality of biomedical research. Because skepticism surrounds science today, we must provide assurances.

Here we will explore QC and QA systems as they exist and develop a general strategy for developing a quality system in the laboratory setting.

Quality systems — QC and QA

QC and QA are distinctly different concepts. However, individuals within quality professions may use the terms interchangeably with no clear distinction. I prefer to divide the concepts by processes. Quality control are standards, processes, and procedures established to control and monitor quality. Quality assurance is the process of reviewing the activities and quality control processes to assure management that the final product meets predetermined standards of quality. There may be some

Table 4.1   Basic Concepts of QC and QA

Quality control systems	Quality assurance systems
Develops standards	Assures that standards are in place
Performs testing of products	Monitors the testing so that proper procedures are followed
Monitors test results against acceptability standards	Inspects test results to assure records are properly maintained and results communicated

redundancy in practice between QC and QA, but the purposes are separate. Table 4.1 shows some of the distinctions.

Differences in QC/QA as required by government agencies

A variety of QC and QA systems are mandated by federal law. Preclinical and clinical experiments that are performed in support of a marketing permit for a food additive, drug, or device must comply with FDA Good Laboratory Practices (GLPs) and Good Clinical Practices (GCPs), regulations that define systems for assuring the quality and integrity of risk assessment data. To use the results in decision making designed to protect the public health, the data must be trustworthy; the FDA asks labs to provide assurances of data quality. The FDA Good Manufacturing Practices (GMPs) assure that pharmaceutical and medical device manufacturers control the quality of their products and fully document the control processes. The Clinical Laboratory Improvement Act (CLIA 88) was established to raise the level of reliability of clinical laboratory tests to an acceptable and consistent standard. Table 4.2 defines some of the differences in approach of the various regulations.

I have not included academic research in Table 4.2 because no formal regulations govern the quality of academic research and it is generally considered self-regulating. Although the FDA was working on a draft Guideline for Academic Research a few years ago, it was never finalized. Generally, the academic community accepts the peer review system and relies heavily on the concept that science is self-correcting. Some universities and some individual departments have policies for internal review of research data and manuscripts. Reputable journals have established procedures for peer review. At the end of the chapter, an optional exercise provides the opportunity for students and researchers to evaluate this system of peer review.

The regulatory requirements cited above are only that — the regulatory requirements. They are not limits to quality control and quality assurance for the particular types of research. They are not the ultimate in quality systems that should be implemented by a research lab. Some research initiatives do not fall under any regulatory authority. But the high standards of scientific quality still apply. The remainder of this chapter provides a strategy for incorporating QC and QA systems into the laboratory, regardless of the regulatory requirements. This strategy does not replace any regulation or specific regulatory requirement. *It is the responsibility of each researcher to be completely familiar with the regulatory requirements of the work being performed and to comply.* Here I present implementation strategies that will conform to the regulatory strategy or will augment it.

Question of quality

What standards of quality should be applied to research? This depends strongly on the nature of the experimental design and what impact the quality of the data has

Table 4.2 QC/QA Regulation in Biomedical Research

Research setting	Regulator and regulation	Quality system	Key elements and practice
Preclinical tox testing	FDA GLP and EPA TSCA and FIFRA GLPs	QC is implied but specify detailed requirements for QA	Management control QAU-QA inspection and audit Study management assurances
Human clinical trials	FDA GCPs	QA system	Institutional Review Board (IRB) Clinical monitors Management assurances
Manufacture of pharmaceutical and medical devices	FDA cGMPs	QC and QA	QC unit tests products Inspects labeling QA reviews recordkeeping
Human clinical testing	CLIA	Accreditation	Internal control procedures Performs blind sample analysis
Academic research	NIH and NSF or other funding agency	Peer review	Funding approval and journal peer review
Patent research	Federal patent laws		

on their intended use. In phase II clinical trials and epidemiologic studies, thousands of subjects are admitted to a study because of the variability of human research data and the inability to control for all variables in the experiment. A large sample size is needed to lessen the criticality of individual data points and to test extremely low incidence levels for rare risk effects. Alternatively, individual test results in a clinical laboratory are highly critical, since one erroneous test result may result in misdiagnosis and even death to a patient. The standard for quality for the individual data will be different, whereas in the case of a clinical lab result, no error outside the testing limits is acceptable.

In toxicological research, all experimental parameters are strictly controlled, from the genetic make-up of the animals to the temperature and humidity of the animal rooms. The only variable is the test substance. Depending on the study design, hundreds to thousands of data points may be collected during the course of a study. What would be the effect on the study if one or more of these were erroneous? This is really a question for statisticians to answer, but here I briefly want to explore the basic potential for any one error leading to misinterpretation of the data. For example, in a dose group containing 50 animals, the probability of seeing a lesion that has a one in a hundred chance of spontaneous development is unlikely; if one of these is missed during the histological exam, the study could be misinterpreted. Alternatively, in the same study, if 1% of body weights out of 1000 weights were incorrectly recorded in the second or third significant digit, because of the natural variability of these weights, there would probably be little impact on the study. However, in the latter example, there are other quality issues beyond the known deficiency in the weights. These issues reflect performance, management control of the study quality, and reliability of other measurements about which doubts could be raised because of the observed flaw in the individual measurements.

The following are fictionalized scenarios to consider, with the data quality criteria preceding, followed by the issues raised by the events.

Were the original observations recorded directly, promptly, legibly in ink?

The errors in the above-mentioned weight data could result from any of these. A technician reads a weight of 267.9 g on the balance, turns to put the animal back in its cage, retrieves the next animal, and hurriedly records 276.9 g on the worksheet. The handwriting is unclear, or the pencil entry is smeared. The data entry person types into the computer spreadsheet 296.9 g. There is now a 29 g discrepancy between the observed data and the reported data. Even though this may not be outside the range of normal variability, other questions arise about the quality of the data. If the technician routinely does not record the data directly, how many other errors have been made? If the handwriting is illegible, what data are lost or misinterpreted during data entry? Should the data entry person try to interpret illegible handwriting? Much deeper is the disregard for recording standards and the necessity for accurate recording of data by the technician and by the data entry person. On a higher level, the standards for acceptable performance and a respect for quality may not have been communicated by study or lab management.

Were the predetermined procedures and methods recorded in the SOPs or protocol followed exactly, and, if not, were any deviations clearly documented in the study record?

The protocol says that all animals will have food and water *ad libitum*. The SOPs detail the procedure for checking, adding, or replacing food and water once daily. An animal care technician enters the room in the morning and notes that one row of animals on a cage rack did not have any food and that the water bottles are missing. From the room log, the last procedure performed in the room the day before was the feeding and watering of the animals. The technician gives food and water to the animals but does not record the error. Because of food and water deprivation, six animals show decreased body weight gain, which is attributed by the study director to a toxic response to the dosing solution. The potential impact on the study of such a deviation is dramatic. The study director could not accurately assess the data because the deviation was not recorded. To carry this example to its extreme, what if the compound were a potential new therapy and was being screened with others to decide which would undergo further experimentation? The results of the experiment eliminated the formulation from further consideration but were based on erroneous information. The technician did not record the error, a deviation from the protocol and standard procedures. Why? Was it to cover up the mistake of the day before? This is fraud. Did the technician get busy with other tasks and forget to report it? It may have happened because of overwork, poor training, or insensitivity to the need for accurate and complete recording of data. These are the result of management flaws and lack of communication, training, and supervision that ultimately affect the quality of data.

Both of these scenarios could have been prevented by the institution of effective quality systems. In the weigh session, a control system may have had one technician handling the animal while the other read and recorded the weight, allowing more time and care be taken in the process. Also double entry or cross-checking of entered data would have caught the data entry error. A QA inspection would identify the flaw in record keeping practice and a QA data and it would catch the data entry error. The food and water fiasco could have been prevented if the lab manager routinely checked all animal rooms at the end of the day. QA inspection of food and water procedures might have prevented the animal technician from not recording the data or at least sensitized the technician to the importance of the data. Finally, a sharp QA auditor might have noticed that all the animals that didn't gain the expected amount of weight were on the same row of the cage rack and reported this suspicious situation to study management.

These scenarios in no way represent all the possibilities for errors that could be introduced into the data. Rather, they simply illustrate the potential for error in all situations and encourage you to develop plans to control and monitor the occurrence of error.

QC and QA approaches

This section discusses QC and QA systems, first defining each system's functions, then providing guidance for implementing each system in a research setting.

The differences between QC and QA

Quality control is planning for the quality of the final product and implementing procedures for testing or inspecting the process against established standards of acceptable quality. The testing and inspecting occurs at various steps in the procedure and evaluates the results or findings against pre-established quality standards. By maintaining quality throughout the development of the product, researchers can then be said to have a known quality.

Quality assurance also involves planning for quality by encouraging the development of quality control processes. Its influence, however, is more enforcement related because additional checks are performed on the procedures, including those of the QC function, assuring management that the resultant product is of the stated quality.

To the end product, QC and QA yield similar results; however, one may not exist without the other if the quality goals are to be met. QC judges quality based on internal standards; QA judges quality based on standards external to the process. Table 4.3 compares QC and QA functions, giving similarities and differences.

Table 4.3 QA/QA Function

Function of QC unit	Function of the QA unit
Defines control procedures that are formalized in SOPs	Inspects procedures performed to determine that the SOPs are being followed; reporting deviations to management
Develops acceptance standards, tests against them, and performs process inspections	Inspects the performance of QC testing and evaluates the procedures against SOPs and other regulatory requirements; reporting deviations to management
Reports test results to management	Audits test results and reports audit findings to management
Personnel within the study group may function as QC	QA personnel must be external and independent of the study group

Implementing a QC program in the lab

Pre-study planning, which is discussed in chapter two means to prethink the processes that a laboratory puts in place to assure the quality of the experimental outcome. This planning is known by different names. In Total Quality Management (TQM) circles, benchmarks are established. Some environmental research groups may refer to pre-study planning as a quality assurance management plan (QAMP) or, on the project-specific level, a quality assurance project plan (QAPP). GLPs do not specify a particular quality planning phase but rather prescribe the elements of management and quality assurance that must be instilled into the study. These involve development of SOPs and of the study-specific protocol. It is expected that the SOPs and protocol will contain sufficient detail, and the procedures for controlling bias will be built into the study. Here I do not wish to champion any particular means to the end. I merely want to explore some of the things to consider when implementing a quality plan. This assumes the lab will have a structure of its own design.

What are the basic necessities in planning for quality? There must be a planner, a plan, and acceptance and implementation of the plan. If there is an organization of any sort, these elements exist. The key is whether they are functioning to accomplish the goal of planning for quality. The planner is management, the organizational unit that has the authority and responsibility for allocation of personnel, funds, and other resources. The plan is how management allocates personnel, funds, and resources. The means for acceptance and implementation of the plan is the strength of the management authority in controlling the activities of the organization.

A research project is to be performed. Management assigns the personnel to perform the work, arranges for the necessary equipment and supplies, and communicates to all involved what is to be done. The effectiveness of management in performing these tasks is directly related to the motivation and context in which decisions are made. If quality is not a key motive in decision making, it is highly unlikely it will be achieved. Unfortunately, quality is often not in balance with other motives, such as profit, meeting deadlines, professional ambition, etc. One way of making sure quality does not get supplanted is to make a conscientious effort to plan for quality. One way to do this is to formalize the planning process and to actively incorporate the topic of quality into the planning phase agenda. Management could then develop in the planning phase: budgeting, personnel assignments, and resource allocations, so that each of these contains the establishment of quality control and quality assurance systems. Plan for QC and QA.

The next level of management planning is the design of the work to be done. In chapter two, there was discussion of the establishment of SOPs. Formalizing this process is essential. Through this process, management communicates the importance of quality concepts such as scientific integrity, accuracy, and consistency. While these concepts are not directly stated in the SOPs, they are inherent in the establishment of standardized procedures and directly relate to the quality criteria for a study. What are these quality criteria? Controlled experimental conditions, consistency in experimental procedures, establishment of standards such as calibration methods, and use of control standards in analytical methods are the quality criteria that are written into SOPs. The following will discuss briefly the basis for each criterion.

Controlled experimental conditions

All factors that could potentially affect the experimental results are controlled to be consistent across experimental groups, except for the substance or procedure that is the object of the experiment. In animal studies, the environmental conditions, husbandry practices, and genetics are some of the conditions controlled while a test chemical is administered at prescribed dose levels. In clinical and epidemiologic studies, the presence of certain health risk factors such as smoking or familial predisposition to disease are controlled. In analytical experiments, the analytical instrumentation and sample storage conditions are controlled to preclude differences in the analysis. For each experimental paradigm there are preplanned controls placed on the procedures to control the quality of the data. This is the establishment of standard operating procedures and the study-specific protocol.

Consistency in experimental procedures

Consistency in experimental procedures is a means for controlling the experimental conditions. Standard operating procedures and the study protocol must be followed consistently to produce unbiased results, results that do not contained unplanned or unknown variables. Quality control and quality assurance procedures are designed to prevent or detect inconsistencies in procedures.

Establishment of standards

Here let us consider what questions are asked to determine what standards to use? The question should perhaps be, "What would happen if a procedure failed or was not performed properly?" Procedures that are critical to the successful completion of the experiment should be evaluated and appropriate check points established, using preestablished acceptance criteria. Equipment calibration, use of untreated or vehicle treated control animals, and analysis of known standards are some of the standards used as a comparison with experimental data to verify experimental results.

There are basically three types of controls or standards. Validation standards are used to test the accuracy of a test paradigm prior to its use in a real situation. Calibration standards are used to assure the functionality of equipment. Verification standards are used to test a system while in use to assure the continued functioning of that system within certain limits of precision.

Which of the controls are most appropriate for a particular experimental procedure depends heavily on what standards are available and the precision required. Is one standardized control sample adequate for an analytical run, or should several be interspersed in the run to check for drift in the calibration? Should a certified standard be used? The answers to these questions must be found in the research methods literature. The above questions were designed to encourage you to evaluate each experimental situation against the concept of controlling for quality.

In-process quality control

Once the plans are made and the experimental work is about to begin, a system of quality checks begin. That is a continual critical review of the experimental processes performed by the technical staff and study management or by a designated QC person (in the case of GMPs). This involves monitoring the control processes, reviewing the calibrations, evaluating standards against tolerances, and in some cases, performing analysis on blind samples. These are all QC functions. In manufacturing settings, a QC unit is established and tests samples of products at varies stages of manufacture. Additionally, the review of records during the experiment is performed. In experimental settings, it is good practice for the technical staff to review the records daily, checking the completeness of entries, making sure handwriting is clear, and that all errors are properly corrected. Remember that it is best for someone other than the data recorder to review the data since it will be easier for another person to spot errors. Lab notebooks may be reviewed and countersigned.

The supervisor of the lab also should review the data frequently. This review should include a check of consistency between technicians, particularly when subjective observations are part of the experimental design. Intra-observer reliability is significant bias that plagues certain types of data. Animal behavioral and clinical sign observations, necropsy findings, and histopathological observations are subject to this kind of error. Several controls may be established to encourage consistency among the technical staff. One is to establish a routine system of double evaluation. In animal behavioral studies, the technician and perhaps the lab manager observe and record a series of observation sessions. The manager then compares the results to see if the technician is making judgments consistent with the standard set by the manager. When several technicians are performing a task, the data often reveal inconsistencies. The manager should review the observations, looking for hints that a technician might be overcalling or undercalling a finding or perhaps misidentifying an effect. This problem may also be identified by having the technician review a set of preserved specimens and comparing the results with established standards. Perhaps the most refined system established has been the pathology peer review system. A group of pathologists reviews the slides from a particular study and compares their findings against that of the study pathologist. A particular diagnosis is then confirmed or changed by consensus opinion.

During the internal QC process, corrections can be made to the procedures used. Equipment that may not be functioning at an optimal level can be adjusted or fixed. Procedures that are not providing consistent results can be revised and tightened. Problems in performance can be identified and corrected. Additional training may be required for technical staff. QC becomes increasingly more important in long-term experiments where a problem left unaddressed could severely reduce the overall reliability of the data.

Certification and accreditation

Certifications and accreditations are offered by a number of professional organizations. Professional certification is offered to individuals, acknowledging that they meet predetermined levels of education and knowledge. Many certifying organizations require periodic recertification that requires the individual to keep current in the field of interest. This is a control mechanism within the discipline to assure a consistent level of competence within the profession.

Laboratories may also be certified. Capability-based certification is based on the organization's ability to perform a particular function and usually involves submission of a report describing management structure, facilities descriptions, personnel qualifications, and other documentation. ISO certification is an example of this type of certification.

Laboratory accreditation is a mechanism for establishing standards for labs performing particular operations. Most accreditations are performance based and usually require demonstration of compliance with standards of organization and method. The accrediting body often inspects the lab at routine intervals to certify compliance. The lab may also be required to analyze proficiency samples periodically to demonstrate the ability to perform according to established proficiency levels.

The processes involved in seeking or maintaining certification or accreditation are often time-consuming and costly. However, there are significant benefits in receiving constructive and unbiased criticism of lab operations. At the end of this chapter I will discuss how to prepare and survive external site visits and audits.

The quality assurance unit

With so many QC checks performed by the lab, what function does QA have? The purpose of QA is to assure management that the controls put in place by QC are functioning to produce a quality product. QA inspects the processes that are critical to the results of the study. QA offers an independent review of these processes. Inspections are performed to assure that the SOPs and protocol are being followed, by observing the laboratory activities in progress, and observing the record keeping practices. Then, at the end of the research project, QA audits the data and the final report for completeness, consistency, and accuracy.

The GLPs require the establishment of what is called the Quality Assurance Unit (QAU). This is a person or group of persons who are separated organizationally from study management. The individuals performing the QA function must be independent of the personnel directly involved in the study to avoid conflicts of interest. This generally means that the QA person reports to a separate or higher level chain of command. The freedom from intimidation and the unbiased support of management are key factors in the success of the QA process. The most common organization of the QAU sees one or more individuals solely committed to the QA function and who report to a high level of management in an organization, often a research vice president. Small companies and universities may have difficulty establishing a QAU because their management structures do not have clear channels of responsibility. Often the QAU is one person who may have responsibilities in other areas. However the unit is developed, the key to compliance is that the individual must have clear independence from the study director and be able to report freely to a management level with responsibility to the study.

The same is true when establishing a QA function in a non-GLP environment. The QAU are watchdogs protecting the legal, ethical, and scientific interests of the organization. There could be situations when whistle-blowing is an appropriate action for the QA person because of his/her knowledge of study activities. While this last statement may cause discomfort in the reader, the solution is to establish responsible management procedures for internal handling of legal or ethical dilemmas, thereby negating the need for QA actions of this type. This discussion was designed to bring the monsters out of the closet. In academic areas, one of the reasons for not instituting QA may be fear of this sort of oversight. It need not be. Again, QA should function to protect the organization from such difficulties. The following discussion concerns how to establish a QA unit that will do just that.

Establishing the QAU

First begin by assigning QA responsibilities to an individual or group of individuals and designating their chain of command. The QAU may be comprised of part-time persons who are also laboratory personnel, as long as they do not review work performed by their lab and do not report on a personnel level to the study director. The QA person must be properly trained in the regulatory requirements of the work, in auditing techniques, and have an ability to understand scientific procedure. This does not mean that the person has to be completely familiar with the discipline of science to be reviewed, but should be able to figure it out with some assistance from the scientist. QA is not peer review. It is a review of procedures and practices in obtaining, recording, and reporting data.

Training in quality assurance is available from a number of different sources. The Society of Quality Assurance based in Alexandria, VA is a professional organization that sponsors basic training and advanced level training at different times during the year. The organization is currently pursuing the establishment of professional certification for QA professionals. A number of private consulting groups also provide training in QA. There are a growing number of books and journals that cover various QA topics. The major authority in QA is the regulations published by the regulatory agencies. The agencies also publish guidance documents that are extremely useful in interpreting the regulations. Because interpretation of the requirements is complex and often confusing, all of these resources must be available to the QA person. One sentence in the regulations may be interpreted in unexpected ways, so that application of the requirements becomes a constant challenge.

Function and goals of the quality assurance unit

The QAU will inspect critical phases of the study and audit the data and report generated from the study. The next sections will discuss both of these functions. The goal of the entire process is to assure management that the studies have been conducted according to applicable regulations, all approved procedures in the SOPs and protocol, and to verify the completeness, consistency, accuracy, and reconstructability of the study records and report. It will be very important for the auditor to remember these criteria. One very strong temptation for the QA auditor is to make a judgment as to the overall "quality of the study" that is based on quality criteria outside the responsibilities of the QAU. If all the above quality criteria are met, then the scientific quality and usefulness of the data can be evaluated by the study director, facility management, the sponsor, and the regulatory authority, not the QAU.

When a GLP study report is finalized, the auditor provides a quality assurance statement for inclusion in the report. This includes a list of the inspections performed and a statement that indicates that the final report accurately reflects the raw data. Essentially, all this says is that the report is a complete, consistent, accurate, and reconstructable representation of the raw data.

Critical phase inspections

QAU periodically inspects the critical phases of a study and assesses compliance with the GLPs and evaluates the consistency between the protocol, the SOPs, and the actual conduct of the study.

The first critical phase is the production and approval of the study protocol. The protocol is critical to the conduct of the study and should be reviewed prior to signature by the study director and sponsor. The protocol should be read in its entirety, noting on the pages any questions, errors, and inconsistencies. This inspection checks for numerical errors, inconsistencies in descriptions of procedures, clarity and completeness of procedural descriptions, citations, and referencing errors. For GLP studies, protocols are required to contain specific information. This should be checked and noted, check also that there are sufficient lines on the protocol for approval signatures required by different regulations. (See Figures 4.1 and 4.2.)

The next step in the inspection process is to identify those phases and procedures of a study that are of critical importance to the quality and integrity of the results. Critical phases for a study should be determined prior to the start of the study with concurrence of the study director. Some examples of critical phases for an *in vivo* assay include animal receipt, quarantine, receipt and storage of test chemical, dosage formulation and analysis, dosing of animals, clinical observations and body weights, and necropsy. For analytical studies, critical phases include chemical or specimen receipt and storage, preparation of samples, performance of analyses (such as IR, NMR, GC, etc.), and facility inspections.

CONFIDENTIAL QA INSPECTION FINDINGS		SOP No.

QA Report No.	Phase	Inspection Date
Study Inspected	Study Director	QA Specialist

QA Inspection Findings	Management/Study Director/PI Response	QA OK'd

For Study Director Use: Please Respond, Sign and Return to the QAU within 5 working days.

Date Received   S.D. Initials   Date of Response S.D. Signature

For QAU Use:

Date Report Sent to Study Director   Date Report Sent to Management   Resolution: QA Signature and Date

[a]QA, ✓ = Finding has been resolved. X = Pending Discussion with Study Director (To be Recorded on Form)

Page ____ of ____

Figure 4.1   Protocol review form.

The GLPs require that inspections be conducted at a frequency adequate to ensure the integrity of a study. This is interpreted to mean that at least one inspection per study is performed. For very short duration studies, one inspection may cover more than one critical phase. For studies lasting a month or so, inspections should be made of the major critical phases. For long-term studies, some critical phases should be inspected at intervals. Chronic toxicology studies may require that certain phases such as dosing, clinical signs, and handling of early deaths be inspected several times over the term of the study. This is to assure that procedures are consistently being followed and that changes to the SOPs or protocol are properly documented and implemented.

QA PROTOCOL INSPECTION

Confidential QA Report No: ________________ Date: ________________

Study Inspected: ________________ Phase: Protocol Review

Study Director or PI: ________________ QA Auditor ________________

Item #	Page #	QA Findings	PI Study Director Response	QA ✓

Page ____ of ____

Figure 4.2 Protocol review form.

For studies of short duration, two methods of inspection may be used. For phases that are one-time occurrences or rarely performed, at least one inspection of an in-process phase (as early in the process as possible) is routinely performed. For studies that are routine in nature and for which the laboratory has demonstrated a proficiency, inspections of all phase may not be necessary for each study but may be held at intervals to assure that the laboratory is continuing to perform according to requirements. The goal of the inspection process is to assure management that the work is performed correctly and that all the regulatory requirements are being followed. Therefore, the inspection process is designed to meet this goal.

Once the critical phases to be inspected have been determined, the inspector will find it necessary to keep track of the laboratory schedule. Whenever possible, inspections should be done at the start of a critical phase in order to insure that the work complies with the protocol the first time the procedures are performed. For each critical phase, identify the procedures to be performed from the protocol laboratory SOPs and regulations. A form may be used to record key words or phrases before the inspection and to note findings during the inspection. Key words are used as prompts for inspection of critical aspects of the procedure.

Performing inspections

To perform the inspection, the QA inspector enters the laboratory wearing the appropriate safety apparel. The QA inspector should bring the appropriate inspection forms (or note pad to record observations). The laboratory should have available the approved protocol and amendments and all applicable SOPs as a reference both for the lab personnel and the inspector. The types of observations made during the inspection will vary according to the nature of the procedure. The inspector should keep in mind the following questions during the inspection:

- Are the activities being performed according to the procedures described in the protocol, the SOPs, and validated methods?
- Are the procedures and data being accurately recorded and documented — promptly, legibly, and in ink?
- Has equipment been properly maintained and calibrated prior to the start of work, and is this properly documented?
- Are reagents properly labeled according to the GLPs and not being used beyond the expiration date?
- Are samples and specimens properly labeled per the GLPs and SOPs?

The observations and findings are be recorded, and a report of the findings is made to the study director. Adverse findings from a QA inspection may vary in their impact on the quality and integrity of the data. Deviations from the GLPs or inspection findings that have a significant impact on the study results may require immediate corrective action. Therefore, these audit results should be reported to the study director immediately.

Should observed deviations be reported to the lab personnel during the inspection? This is a complex issue. Always inform the staff if their safety is in peril or if their immediate actions may result in a loss of data. Otherwise, it may be best practice to allow the study director to review the finding and decide what communication will be made with the technical staff. It is the study director's responsibility to identify appropriate corrective actions, both from a technical and from a personnel perspective.

It is also important to realize that the inspection should not be disruptive of the procedures being performed. From the technicians' view, inspections are uncomfortable experiences. They realize their performance is being monitored and that their work flow may be altered because of the inspector's presence. The inspector should be aware of this. A relaxed and nonthreatening demeanor on the part of the inspector will go a long way to increase the comfort of the technical staff during an inspection.

When findings are reported to the study director (see Figures 4.3 and 4.4), the study director will respond to the inspection report, indicating the corrective actions to be taken. It may then be necessary for the inspector to perform a follow-up inspection on the corrective actions to assure management that the problem has been corrected.

There may be disagreement between the study director and the QA inspector regarding inspection findings or the required remedial action. Procedures should be established for resolving these differences. The first step in all conflict resolution is communication between the parties to insure there is a clear understanding of the

QA INSPECTION KEYWORDS	SOP No.

Confidential QA Report No:_____   Inspection Date: _____

Study Inspected:_____   Phase:_____

Study Director or PI:_____   QA Specialist: _____

General Laboratory Requirements:
❑ Safety Apparel   ❑ Equipment Calibration   ❑ Protocol and SOPs Present

❑ Equipment Logs   ❑ Reagent Labels   ❑ Specimen Labels

Documentation: ❑ Header Information   ❑ Corrections   ❑ Signature and Date of All Entries

Keywords:

_____

_____

_____

_____

_____

_____

_____

_____

_____

_____

_____

_____

_____

_____

_____

_____

_____

_____

_____

Page 1 of ____

Figure 4.3   Critical phase inspection form.

problem. If this does not resolve the problem, then it is the responsibility of the QA person to inform lab management. The GLPs clearly state that management is responsible for ensuring that corrective actions are taken. This allows the QA person to remain independent of the technical conduct of the study.

There is only one instance when the QA person may *not* acquiesce to the final informed decision of management. When fraudulent practices are involved, it is the legal responsibility of the inspector to follow internal procedures and possibly governmental procedures for reporting research fraud.

The inspection process is designed to verify the proper conduct of the study and to provide a means for preventing or correcting activities that could cause the study

CONFIDENTIAL QA INSPECTION FINDINGS		SOP No.

‑ ‑	Protocol Review	
QA Report No.	Phase	Inspection Date

Study Inspected	Study Director	QA Specialist

QA CHECKLIST FOR PROTOCOLS (EPA only requirements italicized)

Yes No

___ ___ (a) Each study have an approved written protocol that clearly indicates the objectives and all methods for the conduct of the study. The protocol shall contain, as applicable, the following information:

___ ___ (1) A descriptive title and statement of the purpose of the study.
___ ___ (2) Identification of the test and control articles by name, chemical abstract number or code number.
___ ___ (3) The name of the sponsor and the name and address of the testing facility at which the study is being conducted.
___ ___ *(4) The proposed experimental start and termination dates.*
___ ___ *(5) Justification for selection of the test system.*
___ ___ (6) The number, body weight range, sex, source of supply, species, strain, substrain, and age of the test system.
___ ___ (7) The procedure for identification of the test system.
___ ___ (8) A description of the experimental design, including the methods for the control bias.
___ ___ (9) A description and/or identification of the diet used in the study as well as solvents, emulsifiers and/or other materials used to solubilize or suspend the test or control articles before mixing with the carrier. The description shall include specifications for acceptable levels of contaminants that are reasonably expected to be present in the dietary materials and are known to be capable of interfering with the purpose or conduct of the study if present at levels greater that established by the specifications.
___ ___ *(10) The route of administration and the reason for its choice.*
___ ___ (11) Each dosage level, expressed in milligrams per kilogram of body weight or other appropriate units, of the test or control article to be administered and the method and frequency of administration.
___ ___ (12) The type and frequency of tests, analyses, and measurements to be made.
___ ___ (13) The records to be maintained.
___ ___ (14) The date of approval of the proposed statistical methods to be used.

___ ___ (b) All changes in or revisions of an approved protocol and the reasons therefore shall be documented, signed by the study director, dated, and maintained with the protocol.

For Study Director Use: Please Respond, Sign and Return to the QAU within 5 working days.

Date Received S.D. Initials	Date of Response S.D. Signature

For QAU Use:

Date Report Sent to Study Director	Date Report Sent to Management	Resolution: QA Signature and Date

Page ___ of ___

Figure 4.4   Critical phase inspection form.

to be invalidated and, therefore, not be useful for its intended purpose. Most often, the QA inspector, the study director, and the study personnel are a team working together to accomplish the goals of the research. The relationships need not be adversarial but rather should be positive and productive.

The data and report audit

There are three stages in the audit process: raw data audit, transformed data audit, and the report audit. The raw data audit involves review of the raw data for completeness and proper record keeping. The audit of transformed data is a check of transcription of raw data to spreadsheets or other data management systems and all summary calculations. This audit follows the data through the data analysis process. The report audit entails checking all sections of the report against the raw data and the transcribed data.

These stages may be audited one at a time, in combination, or all at once. The choice depends on the time allotted to audit and the amount of data or size of the report. For example, large data sets may be audited and the results returned to the lab before data analysis is compete, thus allowing corrections to be made before individual and summary tables are prepared. Next, the raw data is compared to the individual and summary tables, with the results going to the lab prior to the report being written. Finally, the corrected raw data and the individual and summary tables are reviewed in their final form along with verifying the content of the text in the report to the raw data. This is ideal, but requires a great deal of redundancy in handling the raw data. The next best option is to first audit the raw data and the transformed data, reporting the results to the study director so that he/she may use the corrected data to write the text of the report. When the auditor is sure all corrections have been properly made from the first audit, the corrected raw and transformed data can be used to audit the final report.

How to prepare for auditing

Preparation is important. Establishment of criteria for auditing is the first step. Do the data support a GLP study? With what regulatory requirements, including any report format specifications, must the study comply? What data are critical to the outcome of the research? All of these are factors contributing to the development of audit methods and the establishment of the audit criteria.

The objectives of a data and report audit are to:

- Check that the information required by the GLPs is included in the report;
- Confirm that the study was performed in accordance with the approved protocol and that differences in the procedures from the protocol are documented as either an approved amendment or as a protocol deviation (document in the study records and describe in the final report);
- Evaluate the quality and integrity of the raw data according to the principles of the GLPs;
- Check that the report accurately reflects the raw data;
- Check that the report is internally consistent (data in the tables and figures match the text of the report);
- Check for completeness of the study record;
- Check that previous adverse inspection findings have been resolved.

The audit consists of four main steps:

1. Receipt of audit materials: data, study records and draft final report;
2. Audit of data and/or report;
3. Resolution of adverse findings; and completion of remedial actions,
4. Preparation of the QA statement (for GLP studies).

The audit methods should be designed to provide a thorough check of all the study records and the report. This does not necessarily mean that the auditor looks at and cross-checks 100% of the data and report. The auditor may selectively review the data and report and base the selection on the type of data, the method of data collection, the likelihood of errors in certain types of data, etc. The goal of the audit is to identify both random and systematic errors. Random errors are those that occur for no particular reason and usually involve only one data point, e.g., accidental transposition of the digits in a number. Systematic errors involve a set of data points because some systematic error was made. An error in recording a weight or in transcribing a number is a random error. A series of means may be erroneous because the wrong n was used in the calculations and is an example of a systematic error.

The discussion of each audit stage will present in more detail the process of selecting audit samples and procedures for identifying random and systematic errors.

Data audit

Data auditing is the process used to review the records of the study that include the raw data and all supporting documentation. Protocols, SOPs, and regulations should be reviewed prior to starting the audit to familiarize the auditor with the requirements of the study. The goal of the audit is to objectively sort through the data, attempting to reconstruct the activities of the study. A certain unfamiliarity with the data collection system allows the most objective look at the data and their organization and compliance with requirements. This is why independent audits, those performed by sponsors or government inspectors, often detect flaws in the study that internal QA did not identify. It is important for internal auditors to freshen their perspectives on familiar work and take new approaches to the data. This may be accomplished by reviewing data in a new or different order or to track information backwards, e.g., follow data from the results to sample receipt.

Data audits may be performed periodically during a long-term study or at the end of short-term studies. Labs that perform routine analyses may be audited at intervals such as quarterly or bi-yearly.

The auditor begins by reviewing the records that describe the conduct of the study — memos, correspondence, test article and test system receipt, and tracking records. Then all notebooks and raw data forms are reviewed for record keeping practices, completeness of data recording, signature and dates, error correction, and documentation. Next the data record is reviewed for completeness of the study record and for all protocol and SOP required documentation. Finally, the auditor looks for unusual data, outliers, or deviations that may have been the result of unforeseen events during the conduct of the work. Deviations from standard procedures or specified conditions must be fully documented and explained.

Amount of data to audit

Some data sets contain hundreds of data points. It may not be necessary to audit every data point to determine if the data are properly presented. The level of scrutiny depends on several factors. These factors contribute to decisions about whether to perform a 100% data audit or select a percentage of data for review and are based on risk factors inherent in the method of collecting the data. The goal is to review all procedures and experimental endpoints. This is the first level of examination. The auditor should review each procedure performed and all endpoints for compliance with the protocol. The next level is to examine the data for each procedure and endpoint individually. The question whether to review all or part of the data is necessary to determine if the data are accurately recorded.

Basically, small data sets logically receive a 100% data review. That is, all procedures and endpoints and all data are reviewed. Large data sets may receive a percentage review. Here all procedures and endpoints are reviewed, but only a subset of the data is evaluated. For example, 10% of the body weights are reviewed. This is an arbitrary number. The decision of what percentage of data should be audited should be established with consideration of the method of data collection and the sensitivity of the results to errors in the data. Table 4.4 presents factors used to make decisions in how much data should be reviewed.

Table 4.4   Audit Considerations Comparing Data Collection and Risks of Data Loss

Data collection considerations	Risk factors influencing the level of scrutiny
Data collection method	Were the data collected by hand or direct computer capture?
Potential for time-dependent variability	Were the data collected over several days or months?
Training and experience of the technical staff	Intraobserver reliability: did different people collect data?
Supervisory control	How could this impact the data?
How critical is each individual data point?	Were duplicate and triplicate analyses performed?
How are the results analyzed?	Are the data combined in a mean or median?
The number of data points	
What is the sensitivity of the experiment or standard of acceptable variability in the data?	Dose-related trends in pathological data, historical control data

Methods of data collection include hand-recorded data, computerized direct capture of data, and compilation of study documentation. Each method has, inherent in it, a unique potential for inaccuracy, incompleteness, misrepresentation, and inconsistency. The effect on the final results is dependent on the sensitivity of the results to errors in each type of data. For example, hand-recorded data are most susceptible to errors that involve accuracy, completeness, and consistency. A technician weighs an animal and writes down the value displayed by the balance. For that particular weight, the number may have been recorded incorrectly, the handwriting may be illegible, or the balance may not have been zeroed. These are all things that impact only the single data point.

When the same process is performed using direct computer capture of data, there may be no question about the value recorded or the ability to read the number. Perhaps even the zeroing of the balance is controlled by the computer. There are other types of failures that can occur affecting the raw data. The animals may have been weighed in the wrong order, resulting in the recorded value being assigned to the wrong animal. A computer malfunction or power failure could erase data from the file. The program was set up for mouse weights and will reject rat weights entered, not recording them because they are out of a specified range. These failures are systematic and more costly in terms of loss of data. All computerized data collection systems should have extensive performance testing performed prior to their use to collect data. However, there may still be errors in operation and execution of the system that should be evaluated by the auditor.

What effect the errors have on the interpretation of the results depends on the significance of experimental endpoints and also the strength of the response curve and variability inherent in the analytical procedure. This can not be easily generalized for different study designs. The auditor should consult the study director or knowledgeable researcher when considering the effect of certain errors and developing audit methods designed to identify them in particular study types.

Under each method of data collection are different types of data. These include recorded raw data, calculations and procedural descriptions. Again, the amount of scrutiny necessary for these data depends on the collection method as well as the data type. Hand calculations require a 100% check, because of the susceptibility to errors in the formula and input variables. Groups of calculations performed by a computer program can be verified to assure that the formula is correct and the input data are complete. Therefore, a percentage of the data may be evaluated and recalculated.

Selection of data to audit

The above discussion has been in preparation for establishing a systematic approach to auditing. The auditor receives the data to be audited and inventories the material. Then the auditor makes decisions about how much of what data will be checked. Now the question is, how to select the individual data points. There are several approaches that can be used. (1.) Totally random selection uses a matrix of randomly selected areas on a page of data. (2.) Random selection of samples or test systems involves randomly selecting by sample number or animal identification numbers to follow through the study records and across all measurements. This is sometimes referred to as cross-sectional selection. (3.) Random/focused selection may combine random selection with nonrandom selection. For very large data sets, this may be a useful approach. A focused audit selects random samples to evaluate, but selects only certain critical time or measurement sets within those data to review. For example, in a 2-year bioassay, an auditor may select at random a 10% sample of animals, but follow the animals through only the 1st month and the last 6 months of the study for such things as body weights or clinical signs. (4.) Focused selection of data alone is not recommended for most kinds of audits. The auditor selects certain data to review. By doing this, the auditor risks missing critical errors that may affect the accuracy and integrity of the study. If the goals of the audit are more general — are there study records or was a particular parameter measured? — then this approach may be appropriate for a "quick and dirty" review.

Whichever method of selection is used, the auditor's intuition comes into play in the audit. By this I mean, even when a totally random selection is made, it will be important for the auditor to be cognizant of the other data and attuned to recognizing errors in the data not selected. For instance, after checking the sample on a page of animal weights the auditor should scan the other data for something that may draw his/her attention. An experienced auditor develops an ability to pick out potential errors by this method. It is not wise for the auditor to rely too heavily on this approach and abandon the random selection method.

If a random percentage of data is audited, what information does this give to the auditor and the study director about the whole of the data set? The answer to this question can be quite complex. Error rates may be determined, and acceptable standards may be set that prescribe remedial actions based on the error rate. However, for many types of auditing, this information is not useful. Acceptability of errors is not the issue, since all errors identified are unacceptable and must be corrected. The issue becomes whether the corrective action requires a complete reevaluation of all the data or whether it is appropriate to correct only the identified errors. The auditor may make a judgment based on how many errors were found per number of data points audited, whether the errors were of a random nature or systematic, and what effect the errors have on the results of the study. The auditor can make recommendations to the study director with justification for the recommendations based on this assessment of these findings.

Data audit procedures

The following are the procedures used in performing a data audit:

- Gather the study data along with the drafts of individual data tables and summary tables prepared for the final report (if available).
- Inventory the records for any missing data.
- Determine the amount of data to be evaluated.
- Record findings (with reference to their location) on the data audit form and report to the study director (see Figures 4.5 and 4.6).

Hand-recorded data

In most cases, a 100% check of the records for recording errors should be done. Fip through each page to check that all entries are properly recorded, all blanks on forms are completed, and all pages are properly signed and dated. All loose pages should have a unique study designation, a code, or title. Recording errors should have been crossed out with a single line, initialed, dated, and a reason for the change given. All additions to the data, inserted information made after the original entry, should be signed and dated. Were the entries in ink of the proper type and are they legible?

Sample or animal tracking may be performed on a selected percentage of data for large data sets: 10% is commonly used, but this may be adjusted to accommodate unique characteristics of the data. A straight percentage such as 10% of the animals or samples, 10% of the sample time points, or 10% of hand calculated values may be appropriate. These are followed from receipt of the animals or specimen to the final disposition and results. A cross-sectional approach may be taken to certain types of data. The sample should be tracked as already mentioned, but then an additional review is done on a particular parameter or time point. For example, the clinical observations for 10% of the animals are tracked throughout the study; then an additional audit sample of clinical observations is checked for selected days on study. These may be random or selected because they are critical time points, such as the start and end of the data collection time period.

When evaluating calculations performed by a computer and transcription of data in computer data files, it may only be necessary to evaluate the completeness and consistency of data entry and accuracy of calculations, checking for systematic errors. If it is a small data set, plan to review 100% of the data. Otherwise plan to review 10% or more of the main data points. The selection of data points will depend upon the type of data audited and should be based on your experience with the lab and professional judgment. For example, for in-life data, 10% of the animals selected from all dose groups may be chosen for review of all dependent variables (e.g., body weights, food consumption, etc.). For certain types of data it will be necessary to evaluate more of the data because the nature of the data collection contributes to the potential for error. Also, if it becomes evident that no errors or very few are being identified in a particular data point, the QA auditor has the discretion to decide to not do the minimal 10% review. The goal of the review is to establish a confidence in the record that significant undiscovered errors are unlikely.

- Check the data for calculation and transcription errors.
- Review the study records and any record of procedures and methods for completeness, consistency, and reconstructability.

```
┌─────────────────────────────────────────────────────────────────────┐
│                                                                       │
│           QAU DATA AUDIT COVER SHEET            SOP No.                │
│                                                                       │
├───────────────────────────────────────────────────────────────────────┤
│                                                                       │
│   Study Title: _____  │
│                                                                       │
│   _____    │
│                                                                       │
│   _____    │
│                                                                       │
│   Study Code and Protocol No: _____    │
│                                                                       │
│   Study Director: _____    │
│                                                                       │
│   Audit Date(s): _____  Auditor(s): _____     │
│                                                                       │
│   Findings reviewed and a response provided:   Errors/findings have    │
│                                                been corrected or        │
│                                                resolved:                │
│   _____      _____       │
│      Study Director or PI/Date            QA Auditor / Date             │
├───────────────────────────────────────────────────────────────────────┤
│   List of Records and Data Audited: _____    │
│   _____     │
│   _____     │
│   _____     │
│   _____     │
│   _____     │
│   _____     │
│   _____     │
│   _____     │
│   _____     │
│   _____     │
│   _____     │
│   _____     │
│   _____     │
│   _____     │
│   _____     │
│   _____     │
│   _____     │
│   _____     │
│   _____     │
│   _____     │
│   _____     │
│   _____     │
│   _____     │
│                                                                       │
└───────────────────────────────────────────────────────────────────────┘
```

Page ____ of ____

Figure 4.5 Data audit form (page 1).

- Identify clearly findings in the data using a post-it note (do not write on raw data), and describe them in detail on the data audit form.

If an unusually large number of errors are found in the early stages of the audit, the QA auditor may opt to return the whole data set or report to the laboratory for correction. Numerous corrections will require that a second audit of the data be performed resulting in additional expense. Study directors will be encouraged to have an adequate QC program in their laboratories to detect and correct errors prior to the QA audit.

QAU DATA AUDIT FORM		SOP No.

Study Code and Protocol No: _____

QA Specialist: _____

FINDINGS:

Item #	Location and Description	R/✓	QA ✓

✓ = Agree with Finding and Corrected   R = Response Made on Audit Response Form   QA✓ = Finding Resolved

Page ____of ____

Figure 4.6   Data audit form (page 2).

Types of errors in the data audit

The QA specialist performing the audit will look for the following kinds of errors:

Numerical errors: Errors in the number for any of the digits of a datum. Numerical errors occur during the recording or transcription of data. They can often be detected because of an inconsistency between time points or records. Numerical errors may occur during transcription of data from one record to another during data entry into a spreadsheet or data management system. A numerical error can also occur during the recording of the original observation and may be detected if the value is unusually high or low compared to surrounding data points, historical controls, or other study controls.

Calculation errors: Errors in numerical data resulting from an error in a calculation or statistical analyses of the raw data. These may result from an error in the formula used, from data input, or incomplete inclusion of all data.

Recording errors: Errors in the recording of an observation or errors due to a deviation from the specific requirements of the GLPs on recordkeeping. The GLPs specify:

- That data be legible and in ink;
- That data be recorded directly, promptly;
- That all entries be dated on the day of entry and signed or initialed by the person entering the data; and
- That any changes to the data are made so as not to obscure the original entry; and that changes are initialed, dated, and a reason for the change is given. The changes or corrections to the record are initialed and dated by the person making the correction.

Completeness errors: Errors in any study record due to an incomplete entry or entries. Completeness errors occur when:

- Data records are missing (i.e., a data point is required but is not in the records);
- Information requested on preprinted forms is not entered;
- When the specific information required by the GLPs is not included in the study record;
- Records referred to in the audited data set are not available for review; or
- Protocol amendments are incomplete or missing.

GLP deviations: Errors in the study record that occur when there is noncompliance with a specific requirement of the GLPs (other than the GLP requirements mentioned above for recording errors). For example, a "GLP" error occurs when a protocol is not signed by the sponsor.

Report audit

After the data audit has been completed, the final report is audited. The emphasis of the final report audit is to assure that the information required by the GLPs is included and that the final report accurately reflects the raw data and study methods used. The checklist (found on the report audit form, see Figures 4.7, 4.8, and 4.9) is the minimum information required by the GLPs and is used to evaluate the report: [**CAUTION**: EPA requires the same items as FDA, plus additional items. These additional items are listed below in **bold type**.] The information requirements in a final report are similar to the information requirements for the study protocol. Some of the information requirements may not be applicable for a given study (i.e., animal weight range for an *in vitro* study) and can be omitted, but otherwise the above information must be included in the final report.

The following procedures are used in performing a report audit:

- The text portion of the report should be read in its entirety.
- The methods stated in the report are compared to the methods stated in the protocol and the SOPs and the performance is verified in the raw data.

QAU REPORT AUDIT COVER SHEET   SOP No.

Report Title:_____

_____

Study Code: _____ Protocol No.: _____

Study Director: _____

Audit Date(s): _____ Auditor(s): _____

Findings reviewed and a response provided: Errors/findings have been corrected or resolved:

_____   _____
     Study Director or PI / Date   QA Auditor / Date

EPA GLP Compliance Checklist
Reports of Studies

Reference: EPA/GLPs section 160.185 and 792.185

__1) Name of Facility, Address of Facility, Dates Study initiated and completed, terminated, or discontinued.

__2) Objectives Stated in Protocol, Procedures Stated in Protocol, Changes from Original Protocol.

__3) Statistical Methods for Analyzing Data.

__4) Test, control, and reference substances identified by Name, CAS or Code Number, strength, purity, composition, or other characteristics.

__5) Stability of test, control, and reference substances under the conditions of administration.

__6) Description of Methods Used.

__7) Description of Test System Used: (Number of Animals, Sex, Body Weight Range, Source of Supply, Species, Strain/Substrain, Age, Method of Identification).

__8) Description of Dosage, Dosage Regimen, Route of Administration and Duration.

__9) Description of all circumstances that may have affected the quality or integrity of the data.

__10) Name of Study Director, Name of other Scientists or Professionals involved with study, Name of Supervisory Personnel.

__11) Description of the transformation, calculations, or operations performed on the data; Summary and analysis of the data; Statement of the conclusions drawn from the data analyses.

__12) Signed and dated reports from each scientist or other professionals involved in the study, including each person who conducted an analysis or evaluation of the data or specimen after data generation was completed.

__13) Storage location of all specimens, raw data, and Final Reports.

__14) QAU Statement as described in EPA/GLP Section 160.35 (b)(7) or 792.35 (b)(7)
—— (in progress)

Page ____ of ____

Figure 4.7   Report audit form (page 1, FDA).

- The text of the report is evaluated for clarity and accuracy of reporting.
- The numerical data of the tables and graphs in the report are checked against the values found in the summary and individual tables which were previously verified in the raw data during the data audit.
- The internal consistency of the report is then evaluated. The values stated in the text along with the study conclusions are checked against the values found in the tables and graphs.
- All tables and graphs are reviewed for completeness and accuracy of titles, headers, and footnotes.
- The report format is evaluated against protocol, contractual, and regulatory requirements.

```
┌─────────────────────────────────────────────────────────────────────┐
│        QAU REPORT AUDIT COVER SHEET            SOP No.                │
├─────────────────────────────────────────────────────────────────────┤
│                                                                       │
│   Report Title:_____      │
│   _____           │
│                                                                       │
│   Study Code: _____  Protocol No.: _____      │
│   Study Director: _____             │
│   Audit Date(s): _____  Auditor(s): _____      │
│   Findings reviewed and a response provided:   Errors/findings have been corrected or resolved: │
│   _____    _____ │
│        Study Director or PI / Date              QA Auditor / Date      │
└─────────────────────────────────────────────────────────────────────┘
```

FDA GLP Compliance Checklist
Reports of Nonclinical Laboratory Studies

Reference: FDA/GLPs section 58, 185

__1) Name of Facility,
__ Address of Facility,
__ Dates Study Initiated and Completed

__2) Objectives Stated in Protocol,
__ Procedures Stated in Protocol,
__ Changes from Original Protocol.

__3) Statistical Methods for Analyzing Data.

__4) Test and Control Article Identification
(Name, CAS or Code Number, Strength,
Purity, Composition, other
characteristics).

__5) Stability of Test and Control Articles
under conditions of administration.

__6) Description of Methods Used.

__7) Description of Test System Used:
(Number of Animals, Sex, Body Weight
Range, Source of Supply, Species, Strain/
Substrain, Age, Method of
Identification).

__8) Description of Dosage, Dosage Regimen,
__ Route of Administration and Duration.

__9) Description of all circumstances
__ that may have affected the quality
__ or integrity of the data.

__10) Name of Study Director,
__ Name of other Scientists or
__ Professionals involved with study,
__ Name of Supervisory Personnel.

__11) Description of the transformation,
calculations, or operations performed
on the data;
__ Summary and analysis of the data;
__ Statement of the conclusions
drawn from the data analyses.

__12) Signed and dated reports from each
scientist or other professionals
involved in the study.

__13) Storage location of all specimens,
raw data, and Final Reports.

__14) QAU Statement as described in
FDA/GLP Section 58.35 (b)(7)
—— (in progress)

Page ___ of ___

Figure 4.8   Report audit form (page 1, EPA).

- All findings are clearly identified on the report, with a concise explanation, flagged (with a stick-on note), and described in detail on the report audit form (see Figures 4.7 through 4.9).

Findings from the report audit are recorded and are reported to the study director. The forms used to document the findings of the audit are shown in Figures 4.7 through 4.9. Each error, adverse finding, or comment is numbered and recorded on the audit forms. The location of the error is given, and the QA specialist then writes a description of the problem and suggests how it might be corrected. One way to flag errors is to use sticky notes placed on the page of the report with the number of the error identified. The finding may also be described on the page of the report.

QAU REPORT AUDIT FORM				SOP No.	

Study Code and Protocol No: _____

QA Specialist: _____

FINDINGS:

Item #	Loc of Error	Description	R/✓	QA ✓

✓ = Agree with Finding and Corrected. R = Response Made on Audit Response Form QA✓ = Finding Resolved

Page ____ of ____

Figure 4.9   Report audit form (page 2).

The completed QA data and report audit forms are sent to the study director for review and comment. After the study director has completed the review, the audit and corrected data and report are returned to the QA auditor for review of the changes. When the audit has identified significant problems, a second audit of the corrected report or data set may be necessary to assure compliance with the GLPs.

A QA statement is required to be included in all GLP final reports. It is a signed report by the QA inspector or auditor of the inspections performed and includes a statement that the final report accurately reflects the raw data. Each of the QA findings should be resolved and any corrective actions completed prior to the auditor signing of the QA statement. The QA statement is a specific requirement of the GLPs and should include the following information:

- Study title and other identifiers of the study.
- The date inspections were made.
- The date the inspection findings were sent to the study director.
- The date the inspection findings were sent to the management.
- The signature and date of the QA person.

Other optional information that is often included on QA statement is

- the critical phases inspected
- the dates of data and report audits
- a statement indicating that the report accurately reflects the raw data

Upon completion, the QA statement is given to the study director for inclusion in the final report. If, after the signing of the QA statement, there are any revisions to the report or changes to the study record that have a significant impact upon either the interpretation of the results or the accuracy of the report, then a second QA audit may be performed. An additional line may be added at the end of the original QA statement (i.e., below the signature) stating that a second audit was performed due to further revisions, who performed the audit, and the date when the audit was performed. The QA auditor performing the second audit then signs and dates the appended QA statement.

Outside inspections

There may be others who will perform inspections on research projects. In a contract research organization, often sponsors of studies will ask to inspect the work in progress or to review the study records. These inspections may be very superficial or quite detailed, something akin to the "inspection from hell". Labs subject to these site visits will find it useful to develop an SOP that details the procedures to be followed by the lab and by the visitors. The SOP can set the stage for a manageable experience.

Labs that are certified or accredited by some body will be inspected in some routine manner. These inspections are often stressful for the staff and management. Again, an SOP that plans for pre-inspection, inspection, and postinspection activities is very useful. The key to surviving these inspections is to maintain the lab consistent with the requirements of the inspectors at all times. This prevents several things, the disruption of normal lab activities by massive preparation (clean-up) efforts, the likelihood someone will lapse into old habits during the inspection, and the communication to the staff that the requirements are unnecessary and burdensome. Also, inspectors can sense when the lab is not functioning normally, which causes them to investigate their suspicions.

Government inspectors or investigators may visit your facility. FDA, EPA, and USDA inspect labs performing work under their regulations. How to survive one of these inspections has been the subject of books. Because the regulatory arena is in constant change, it is important to keep abreast of the changes and of the initiatives of each agency. The Freedom of Information Act allows you to request from the agencies information, in particular their SOPs or procedures for performing inspections. Also there are several private and commercial watchdog groups that disseminate information about the current practices of the inspectors.

One of the most reasonable FDA investigators I can remember told a colleague that their primary objective in the audit was to try to discern how much control study management had and that this could be accomplished in perhaps a day or so at the facility. Control of the study conduct is demonstrated through the documentation record and through attitudes communicated during the inspection. The current approach by the FDA is to look for documentation of management's role in the studies and to document all findings, however small, as a means for constructive input.

The consequences of these audits vary. FDA may require corrective action be taken or ask for voluntary corrective action. In addition, the study or studies from the lab may be rejected, based on the audit findings. The laboratory may be banned from any future work to be submitted to the agency. And, if fraud is found, the lab management and personnel may be charged with criminal mail fraud, sending false documents through the mail. The EPA has adopted a more formal process for legal prosecution of violations. It has established civil and criminal penalties for GLP violations that relate to the severity of the crime and the length of time. A recent case, mentioned in chapter one, resulted in prison terms and fines for management, for lab personnel, and the QA manager. All but management received suspended sentences.

The USDA inspects animal facilities for compliance with NIH Guidelines for the Care and Use of Laboratory Animals. It may cite a facility for violations requiring remedial action or may close a facility if violations are severe.

Other governmental organizations that may inspect facilities are NIDA, DOT, NRC, and various state agencies. Fire and safety inspections also are performed, depending on local, state, and federal laws.

Conclusion

This chapter has discussed two quality systems, quality control and quality assurance. The quality control system establishes the mechanisms by which quality can be built into the study either by a format QC process or by instilling quality checks into the SOPs.

The quality assurance system begins by reviewing the planning to assure management that the quality control systems are in place and that the study plan is complete and is an accurate representation of the work to be done. Then QA follows the conduct of the study observing the QC system and the performance of the work to assure and reassure management that the highest quality standards and regulatory requirements are met in the final product, the report.

Practical applications

Application

For those working in a laboratory, make an agreement with another lab to work jointly on reviewing each other's processes, data, and reports. Do this in such a way and with the attitude that the findings will be used both to improve each other's QC systems and to learn more about the QA perspective.

Alternate application

Ask the QAU if you could work with them on a study to learn audit techniques. This will assist lab personnel in being more attuned to finding and correcting errors.

References

DeWoskin, R. S. (1995), *Quality Assurance SOPs for GLP Compliance*, Interpharm Press, Buffalo Grove, IL.

Drug Information Association, *Computerized Data Systems for Non-Clinical Safety Assessments: Current Concepts and Quality Assurance*, Maple Glen, PA, September 1988.

Guide to the Care and Use of Laboratory Animals, National Institutes of Health, Bethesda, MD, 1985–1986.

Institutional Animal Care and Use Committee Guidebook, NIH Pub. No. 92-3415. U.S. Department of Health and Human Services, pp. E33–37.

Sheenhan, D. C. and Hrapchak, B. B. (1980), *Theory and Practice of Histotechnology*, C.V. Mosby, St. Louis.

Taulbee, S. M. and DeWoskin, R. S. (1993), *Taulbee's Pocket Companion: U.S. FDA and DPA GLPs in Parallel*, Interpharm Press, Buffalo Grove, IL.

U.S. EPA, FIFRA Good Laboratory Practice Standards, Final Rule, *Fed. Reg.* 54:34052–34074, August 17, 1989.

U.S. EPA, TSCA Good Laboratory Practice Standards, Final Rule, *Fed. Reg.* 52:48933–48946, August 17, 1989.

U.S. FDA, Good Laboratory Practice Regulations, Final Rule, *Fed. Reg.* 52:33768–33782, September 4, 1987.

U.S. EPA (1986), Pr Notice 86-5, U.S. EPA, Washington, D.C.

U.S. EPA (1995), Good Automated Laboratory Practices, Office of Information Resources, Research Triangle Park, NC, August.

U.S. FDA, Good Clinical Practices: *CFR*, Title 21, Part 50, 56, 312, 314, April 1, 1993.

U.S. EPA, Enforcement Response Policy for the Federal Insecticide, Fungicide, and Rodenticide Good Laboratory Practices (GLP) Pesticide Enforcement Branch, Office of Compliance Monitoring, Office of Pesticides and Toxic Substances, U.S. EPA.

U.S. FDA, Current Good Manufacturing Practices for Finished Pharmaceuticals, *CFR*, Title 21, Part 211.

U.S. FDA, Current Good Manufacturing Practices for Medical Devices, General, *CFR*, Title 21, Part 820.

U.S. FDA, Guide for Detecting Fraud in Bioresearch Monitoring Inspections, Office of Regulatory Affairs, U.S. FDA, April 1993.

U.S. EPA, Clean Air Act, *Fed. Reg.* 59, No. 122, June 27, 1994.

U.S. FDA, Electronic Signatures; Electronic Records; Proposed Rule, *Fed. Reg.* 59:13200, August 31, 1994.

Webster's Ninth New Collegiate Dictionary (1983), Merriam-Webster, Springfield, MA.

chapter five

Training

Shayne C. Gad

All the preceding chapters have attempted to deliver an understanding of why proper record keeping is important and how to actually go about creating and maintaining the required documentation. What hasn't been addressed yet is how to train those who need to know such things in these necessary aspects.

Clearly the principal investigator or supervisor in charge of a laboratory is the individual ultimately responsible for insuring the integrity of the research process and of the data resulting from it. These same individuals must therefore see that those working in the lab (or research operation that they are responsible for) are trained, and that such training is adequate, documented, and kept current. And perhaps the best way to consider how to adequately discharge this responsibility is one step at a time.

Who to train

Any and all individuals who are (or are to be) involved in the generation and handling of research data need to receive appropriate training. For those who are already in a lab setting in which this will represent a new or changed practice, a couple of hours of instruction followed by access to appropriate reference materials and clear resolution of any questions should serve the purpose nicely.

Clearly a part of the process of bringing in a new research worker (be it technician, visiting scientist, or student) should be ensuring that he/she is knowledgeable of the accepted standards for data handling in the laboratory. Definitely all new graduate and postdoctoral students should receive a "short course" on the subject when they are first entered into a program of study and research.

When

If the research establishment has been following an adequate set of procedures for the recording of data and methods and for the handling and retention of such records, then the issue of **when** applies only to new members of the research team. Such new team members should be trained to the point that the principal investigator or research supervisor is comfortable with their practices before they are allowed to conduct research activities on their own.

If the following of the principles of documentation and data handling presented in this text represents a significant change from existing practice, however, then a time must quickly be identified for the training of all personnel. Each laboratory has its own pattern of work, and clearly a time should be selected that both causes the least disruption to the work flow and also allows the research team members to be effectively trained. But it should be done as expeditiously as posssible. The National Science Foundation has already started conducting random inspections of laboratories utilizing their grant funds (Anderson, 1993), and similar inspections should be expected from other agencies.

By whom

Actual instruction of new researchers should be by an experienced researcher who is actually currently active in day-to-day laboratory operations. This may mean that a senior technician or postdoc may be the most effective individual when it comes to actual hands-on training. In such a case, however, the principal investigator or supervisor responsible for the reliability of the data and publications coming from a laboratory's work should oversee the training and participate in it to at least some extent.

How

Training should utilize both a lecture and a hands-on approach. The lecture portion needs to cover the reasons for, importance of, and basic aspects of data recording, handling, and maintenance. It should also include explicit instructions as to procedures for when something goes wrong or an error occurs.

The practical portion of the training should be focused on the local situation and particulars. These should include the location of supplies and of stored data, the types of documentation maintained in the lab (including any in place to insure instrument calibration and control of use of test materials), and any special procedures for making copies of data.

Points to consider

There are five essential points which should be addressed in any training given new reseachers in any biomedical research facility. Each and every individual entering into such a work capacity should clearly understand principles, policies, and procedures for working in the research capacity and how they apply for that specific facility. Having a set of concise and clear written guidelines for each such individual as they enter the training "program" is strongly recommended. The specific principles which must be addressed are as follows:

Oversight: There is a specific individual who has ultimate responsibility for the quality of the data produced by the research facility or group or laboratory, and the documentation which supports that quality. The responsible individual is concerned about both data and documentation quality — if it isn't adequately documented, it cannot be considered to have happened for purposes of science. And that concerned individual will spend the time to oversee the work of each of his or her charges to ensure that the quality of both is maintained.

Double copy: At some regular and set interval, an exact copy of all the original data generated by each individual working in the laboratory must be stored in a separate and secure place. For graduate students, the use of carbon papered, numbered page notebooks enforces this habit and should be strongly encouraged. No data that are important should be taken out of the immediate research environment unless an exact copy has been put into a designated safe storage area — an "archive."

Review: At some regular interval (usually weekly), a supervisor or peer researcher should review all original data to ensure that they are clear to another party than the investigator, and should denote that such a review occurred by initialing and dating the bottom of each page or piece of data. If there is any question of clarity, it should be cleared up at the time of the regular review.

Signature and date: A researcher's only real product is her or his data. This is the individual's unique intellectual contribution to society. Each day's entry or page or piece of data should be signed and dated by the individual. In so doing, the individual is signifying that the entry is complete, accurate, and his or her own work.

Secure storage: On some regular and preestablished basis, data must be stored in some place where they are both physically and intellectually secure. The location of the storage facility and of specific materials within it should be well defined and documented. In GLP regulated research, such a place is called an archive. It is the responsibility of the investigator to ensure that such storage occurs.

References

Anderson, C. (1993), NSF's new random inspections draw fire, *Science*, 261:289.

appendix one

Rules for form design and preparation

1. Forms should be used when some form of repetitive data must be collected. They may be either paper or electronic.
2. If only a few (two or three) pieces of data are to be collected, they should be entered into a notebook and not onto a form. This assumes that the few pieces are not a daily event, with the aggregate total of weeks/months/years ending up as lots of data to be pooled for analysis.
3. Forms should be self-contained, but should not try to repeat the content of SOPs or method descriptions.
4. Column headings on forms should always specify the units of measurement and other details of entries to be made. The form should be arranged so that sequential entries proceed down a page, not across. Each column should be clearly labeled with a heading that identifies what is to be entered in the column. Any fixed part of entries (such as °C) should be in the column header.
5. Columns should be arranged from left to right so that there is a logical sequential order to the contents of an entry as it is made. An example would be date/time/animal number/body weight/name of the recorder. The last item for each entry should be the name or unique initials of the individual who made the data entry.
6. Standard conditions that apply to all the data elements to be recorded on a form or the columns of the form should be listed as footnotes at the bottom of the form.
7. Entries of data on the form should not use more digits than are appropriate for the precision of the data being recorded.
8. Each form should be clearly titled to indicate its purpose and use. If multiple types of forms are being used, each should have a unique title or number.
9. Before designing the form, carefully consider the purpose for which it is intended. What data will be collected, how often, with what instrument, and by whom? Each of these considerations should be reflected in some manner on the form.
10. Those things which are common/standard for all entries on the form should be stated as such once. These could include such things as instrument used, scale of measurement (°C, F, or K), or the location where the recording is made.

Index